Praise for *Why Study Mathematics?*

Why Study Mathematics? is an insightful guide for anyone considering studying mathematics at university. It explains the sort of maths you can expect to find and how it will be taught, and highlights the wide variety of career options that a maths degree opens up. It also includes important examples of where maths is used in the real world. I recommend it to all prospective maths students and their parents.

Nicholas J. Higham, Royal Society Research Professor and Richardson Professor of Applied Mathematics, University of Manchester

The mystery of a mathematics degree – what it is, where it leads and why it's useful – is unlocked in this easy read. Detailed, accessible and broad ranging, Vicky Neale refines the complex and varying nature of high-level mathematics into an understandable, useful and relatable form. An ideal guide for A level maths students when pondering their next steps.

Kerry Burnham, headteacher, Exeter Maths School

Why Study Mathematics? explores in depth the various options that a maths degree has to offer as well as providing expert guidance on what to expect from a maths degree. Vicky Neale addresses all the questions that an enthusiastic mathematician considering a maths degree might have. She refutes the myth that studying maths beyond school/college is restrictive or overwhelmingly challenging. Neale's book is both informative and engaging: it made me want to study a mathematics degree all over again, or at least revisit certain topics that are discussed. I highly recommend this book to any student or maths enthusiast wanting to study mathematics at university. Being a mathematics teacher myself, this book is definitely a resource I shall be directing A level students to.

Rebecca Blazewicz, mathematics teacher, Bristol Grammar School

An essential read for anyone considering studying mathematics at university. Vicky Neale takes you through what to expect in your studies, and explains the practical uses and beauty of a mathematics degree.

James Grime, lecturer, public speaker and Numberphile presenter

Another great book from Vicky Neale. An extremely useful guide for students (and their advisors!) on studying mathematics at university.

Dr Kevin Houston, School of Mathematics, University of Leeds

Why Study Mathematics? is superb! It will be the very first book that I will recommend to students who wish to study mathematics at university, as it provides a very easy-to-follow guide to the many fascinating areas and branches of mathematics that could be studied during a mathematics degree and to the ever-growing number of careers that require one. Neale provides compelling insight not just into how useful mathematics is in today's modern world, but also into what an essential and integral element of everyday life the subject is. The book provides wonderful examples of how mathematics is used in today's ever-evolving world, from its involvement in online shopping and JPEG compression to the way it's used to analyse climate change. The beauty of mathematics is also revealed through Escher's paintings, many elegant equations and the intriguing world of infinities, all of which should inspire students to go and find out more.

Jason Hudson, Director of Mathematics, Wilson's School

This book is impressively thorough in its treatment of the factors that a student might consider, both in navigating the intimidating variation of choice in mathematical courses and in exploring the essence of what maths is and the subject's implications in a wide variety of industrial and research settings. Vicky Neale is clearly someone who not only has a deep knowledge of maths, from SIR models to JPEG compression, but also has extensive experience in helping students both in the lead up to university and beyond. I highly recommend it to any student, including both those who are just exploring their university options and those who are already set on a mathematics degree and want to explore the short-term and longer-term implications of their choice.

Dr Jamie Frost, teacher at Tiffin School, founder of DrFrostMaths and Top 10 Finalist for the Global Teacher Prize 2020

Why Study Mathematics? is awesome! It's absolutely the book I wish I'd had as a sixth former, when, truth be told, you don't really know what studying a maths degree is. It's beautifully written and really engaging, and it represents a great starting point from which to explore all the different maths courses that are available. Crucially, it helps you formulate the questions you need to ask the tutors on those different courses to help you work out where will be the best place for you to study.

Sophie Carr, founder of Bays Consulting and Aperiodical's 'World's Most Interesting Mathematician'

This book is essential reading for A level students who are thinking about a maths degree. The reader is given an accurate picture of what to expect on a maths degree course and what employment opportunities may follow. The descriptions and examples are perfectly chosen and do a great job of showing why the subject is so interesting and enjoyable.

David Ireland, maths teacher, Heyford Park Free School

WHY STUDY MATHEMATICS?

The *Why Study* Series

Studying any subject at degree level is an investment in the future that involves significant cost. Now more than ever, students and their parents need to weigh up the potential benefits of university courses. That's where the *Why Study* series comes in. This series of books, aimed at students, parents and teachers, explains in practical terms the range and scope of an academic subject at university level and where it can lead in terms of careers or further study. Each book sets out to enthuse the reader about its subject and answer the crucial questions that a college prospectus does not.

Published

Why Study History? — Marcus Collins and Peter N. Stearns
Why Study Mathematics? — Vicky Neale
Why Study Geography? — Alan Parkinson

Forthcoming

Why Study Languages? — Gabrielle Hogan-Brun

WHY STUDY MATHEMATICS?

BY VICKY NEALE

Published by London Publishing Partnership
www.londonpublishingpartnership.co.uk

ISBN: 978-1-913019-11-2 (pbk)

A catalogue record for this book is
available from the British Library

This book has been composed in
Kepler Std

Copy-edited and typeset by
T&T Productions Ltd, London
www.tandtproductions.com

Printed and bound in Great Britain
by Hobbs the Printers Ltd

Cover image

*Mathematicians play an important role in understanding
and tackling climate change. Mathematical modelling and
statistical analysis allow researchers to review data, to make
predictions about the melting of the polar ice caps and to study
the population dynamics of animals such as polar bears (hence
our cover image). Mathematical expertise is also relevant for
managing the impact of climate change in the energy, health
care, transport, agriculture, retail and finance sectors, to name
just a few. Some of these topics are addressed in this book, and
you'll see a polar bear in a completely unrelated context too.*

CONTENTS

ACKNOWLEDGEMENTS

I'm very grateful to the friends and colleagues who read drafts and offered me wise advice and constructive suggestions: Nathan Barker, Sue Cubbon, Charlie Gilderdale, Mareli Grady, Fiona Hamey, Lizzie Kimber, Derek Moulton, James Munro, Peter Rowlett and Andi Wang. I'm also very grateful to the team at London Publishing Partnership for their expertise and patience – Richard Baggaley, Sam Clark and Ellen White – as well as two anonymous reviewers.

INTRODUCTION

Why study mathematics?

MATHS IS A VERSATILE SUBJECT, with different flavours that appeal to different people with different tastes. It equips graduates with skills that employers value. It's full of fascinating ideas and powerful applications, and the process of understanding a new mathematical concept or solving a problem using maths is enormously satisfying. Whatever your priorities – whether you're looking to help other people, to earn a lot, to explore a creative subject or to make a difference in society – maths has something to offer you. The study of maths is rewarding in and of itself, and it gives you lots of options for the future.

There are many factors to consider when choosing *what* to study at university and *where* to study it (and this follows a careful decision about whether your next step is to attend university or to pursue another path). You might already know from your experience at school or college that you have a particular interest or strength in a certain subject, or you might be choosing between a few options. Or, as some subjects are available at degree level but not at school or college, you might be researching these as well as looking into how subjects that you've previously studied develop and change at university. (As you'll see in this book, maths at university includes a broad range of topics, many of which you won't have encountered at school.) In addition, you might have a particular career path in mind, or you might be looking to ensure that you keep your job options open.

This book will help you to find out more about maths at university. In Part I, we'll explore the practicalities of a maths degree. What's involved in studying a maths degree? What topics might you study? What teaching methods and types of assessment might you encounter? How do you choose between the wide variety of maths degree courses on offer? What makes a good maths student? What careers are open to maths graduates?

In Part II, we'll look more closely at some of the topics you might study at university, providing a taste of the theoretical underpinnings of maths and offering insight into its diverse applications: in medicine and health care, in digital communication, in engineering, in tackling climate change, and more. My choice of subjects is inevitably centred on mathematical nuggets that *I* find fascinating, and I've tried to pick examples of topics and applications that don't often come up at school or college in order to give you a glimpse of further horizons, rather than to remind you of things you already know well. Don't worry if you don't follow all the mathematical ideas – these are topics you might meet in a maths degree, not material you're supposed to understand already!

I've included some suggestions for further reading at the end of this book. These include books that you might want to read before starting a maths degree as well as websites to inspire you and to help with your decision making. There's a lot out there for aspiring mathematicians, from engaging YouTube videos by Numberphile to stimulating maths problems by NRICH and the UK Mathematics Trust, from case studies by Maths Careers to biographies of mathematicians throughout history by MacTutor.

Perhaps you're reading this book as a student looking to make a decision about university. Or maybe you're reading it because you're supporting a family member making this decision, or because you're a teacher working with students trying to choose a university course. Whatever your situation, I hope that this book will give you a clearer picture of why a maths degree is a good option for many people.

If you're the one who's thinking about embarking on this adventure, then I would like to wish you all the very best with your mathematical studies.

Last-minute note

I'm putting the finishing touches to this book in the spring and summer of 2020, under 'lockdown' in the UK because of the Covid-19 pandemic. As you'll see in a few pages' time, Chapter 5 starts with a discussion of the use of maths to study the spread of disease. Believe it or not, this was always the plan for this chapter, long before Covid-19 emerged, and the first draft now looks uncomfortably prescient.

The vital role that mathematical modelling has to play both in predicting how this disease might unfold and in simulating the effect of different strategies is being featured on the national news, along with discussions about 'flattening the curve' and the reliability of statistical data on cases so far.

In addition to modelling the spread of this disease, maths graduates have been developing software and planning logistics for the National Health Service (NHS) and supermarkets; teaching and safeguarding young people; analysing and managing risk in all areas of business; keeping the finance sector open; and supporting the economy, all of which are making a powerful difference during this pandemic.

The world around me as I write is strange and unfamiliar. I'm not allowed to leave my house, except to take a daily walk or to buy groceries. My students are scattered around the world, most having left Oxford at the end of term, just before lockdown was introduced. They're doing an extraordinary job of continuing to study maths and make excellent progress under difficult circumstances. I teach them from my kitchen table, using technology to bridge geographical gaps that now seem very wide.

No doubt you're wondering what this means for the future of university study and specifically of maths degrees. While university planning is well underway for the 2020/21 academic year, we don't really know what this will look like yet. I'm certain that it will have to involve a mixture of online and face-to-face teaching and learning. I hope that, in the long term, we'll take the best aspects of the former and adapt them to improve our maths degrees in the future, using technology to enhance our lessons, but not to replace those activities best done in person. Maths lecturers from different universities are already coming together to exchange ideas about how best to organize teaching and to support students' learning in light of our changing circumstances.

While all university courses might look a little different over the next few years, maths degrees will still offer a stimulating, inspiring, satisfying and rewarding programme of study as well as a great platform on which to build a fulfilling career.

PART I

THE INS AND OUTS OF A MATHS DEGREE

CHAPTER 1

What's in a maths degree?

IT'S A REAL STRENGTH OF the higher education sector that there are so many different maths degrees available. This means there's a good chance you'll find one that appeals to you, and it's worth doing some research to explore the options that might suit you best. In our first two chapters, we'll look at the structure of a maths degree. What topics might you study at university? What kinds of programmes are available, and what might fit your interests? We'll also give you some idea of what it's like to study maths at university, from admissions requirements to teaching and assessment and support services.

What do we mean by 'maths'?

Throughout this book, 'maths' is short for 'mathematical sciences'. The mathematical sciences include mathematics of all sorts (including pure mathematics and applied mathematics) as well as statistics and operational research (which is often shortened to OR). Some universities offer degree programmes that focus on particular aspects of the mathematical sciences, while others offer broader degree programmes that nonetheless give scope for specialization at a later stage. For the purposes of this book, these all fall under the umbrella of 'mathematical sciences', or just 'maths', degrees.

The names of degree programmes vary significantly. On the one hand, two programmes that are simply called Mathematics can have very different structures and content. On the other hand, programmes with different names, such as Financial Mathematics and Mathematics with Statistics for Finance, can have similar content. Some programmes have a lot of flexibility, so a student on a mathematics programme might be able to study a lot of the same content as a student taking financial mathematics, for example.

You'll find that degree programmes with names like Business Analytics might overlap with programmes such as Mathematics,

Operational Research, Statistics and Economics. So it's important to look beyond the programme title (and the UCAS course code) to find out more about what's involved. This book will help you to make sense of some of the topic names you might see mentioned in prospectuses and on the UCAS website (https://www.ucas.com/). This website features a useful glossary of terms as well as loads of information for those contemplating university.

Something old, something new

Some topics in a maths degree will be familiar to you from school or college. They'll build on ideas you've already studied but take them further through the development of new techniques and the discovery of new applications. Other topics will be completely unknown to you. Exactly which topics fall into each of these categories will depend on what you've covered at school; even just within the UK, students take a variety of qualifications (A levels, Scottish Highers, International Baccalaureate, ...).

Maths is a cumulative subject, where being able to study more complex material is dependent on your having grasped earlier topics. There are some mathematical ideas that everybody must study and understand so that they can progress to further ideas and techniques. You've already experienced this concept: for example, you needed to be comfortable using algebra to solve linear and quadratic equations in order to tackle other aspects of mathematics. The cumulative nature of this subject means that maths degrees usually include some compulsory modules on core topics. This enables later modules to assume that everyone is familiar with a certain amount of the material already.

As some subjects included in a maths degree are new and unfamiliar based on school experience, it can be hard to make informed choices about what to study at the start of a degree. These compulsory modules will help you not only to develop a secure foundation on which to build later modules, but also to explore your own mathematical interests, so you can make decisions about what to study later on when you have a choice of modules.

In some degree programmes, you'll choose a large proportion of your modules, especially in later years, while in other programmes most of your modules will be core, with a smaller optional component.

What's your flavour?

At school, there are standard requirements about what students have to learn: there's a national curriculum. This isn't the case at university. Universities have significant flexibility when it comes to how they organize degree programmes. There's a huge variety of courses – offering differences in mathematical content and emphasis, in teaching style and in assessment methods – that can lead you to a maths degree. This means that it's really important to research the available options before you apply, to find courses that'll suit you.

The Quality Assurance Agency (QAA), which is responsible for overseeing standards in higher education in the UK, publishes a benchmark statement for each subject area, setting out expectations for degrees in that discipline. For mathematics, statistics and operational research, it says:[1]

Some courses are concerned more with the underlying theory of the subject and the way in which this establishes general propositions leading to methods and techniques which can then be applied to other areas of the subject. Other courses are more concerned with understanding and applying mathematical results, methods and techniques to many parts of the overall subject area.

The benchmark statement refers to these as 'theory-based courses' and 'practice-based courses', respectively. It continues:

> *While there are a few courses that are entirely theory or practice based, most have elements of both approaches and there is a complete spectrum of courses covering the range between the two extremes. It is possible for courses with the same title to have very different emphases; it is the curriculum of a course (rather than its title) that makes clear its position within the spectrum. It is important to note that all of these different emphases are valuable, and one should not be viewed as of higher status than another.*

There's a lot of flexibility about which topics are covered in a maths degree. The only topics specifically mentioned in the QAA benchmark statement are calculus and linear algebra (see Chapter 5).[2] Beyond that, it's up to individual universities to design appropriate degree courses.

Different people have different mathematical tastes. Some like nothing more than to get their hands on a large data set, to interrogate it in order to see what conclusions they can draw, and to consider the robustness of those conclusions. Others are motivated by a particular application and spend time exploring which mathematical tools can help to answer the questions they find exciting in that area. Others still are fascinated by the beautiful, fundamental questions in this subject (it's surprising just how many fundamental questions there are for which we still don't have complete answers) and devote themselves to curiosity-driven maths.

Within a maths degree, you're likely to have opportunities to experience all of these facets. However, degree programmes do differ in the emphasis they place on each aspect of maths, as well as in the number of options they offer to students, so it's really worth thinking about what style of course will suit you. Are you motivated by the use of maths in industry and other applications, and therefore interested in building a mathematical toolkit for that purpose, or would you relish delving into the background theory of how and why the tools work? You're not restricted to one or the other: many courses combine elements of both. But when you're researching courses, it can be helpful to consider the extent to which each might be described as 'theory-based' or 'practice-based', because viewing courses through that lens might help you to focus on the ones that'll suit you.

The main areas of mathematics

So what kinds of topic might you study at university as part of a maths degree? These might roughly be grouped under the following headings:

- ▶ pure mathematics;

- ▶ applied mathematics;

- ▶ statistics;

- ▶ operational research (OR).

These divisions are necessarily slightly contrived, as there are lots of areas of overlap. For example, you might have already studied some calculus (differentiation and integration) at school or college: this is a good example of a topic that can be approached from both a pure mathematics perspective and an applied mathematics one. Nevertheless, these four headings can still be a helpful starting point for thinking about university-level maths topics. We'll meet a selection of subjects under these headings in Part II. To give you some idea of what they involve, we'll briefly describe each of them in turn below. Many of these topics aren't introduced before university level, and in some cases you won't meet them until the later years of a maths degree; so if the names are unfamiliar to you, or you don't completely understand the descriptions, or you don't know how you'd choose between topics, there's no need to worry! The descriptions are here just to give you a sense of the kinds of topic you might find in a maths degree.

Pure mathematics

Pure mathematics tends to refer to maths that's done for its own sake, rather than with a specific application in mind. It's sometimes called fundamental mathematics and has a focus on rigorous proof: developing careful arguments that show with absolute certainty that theorems are true. It involves precise definitions of concepts so that mathematicians can reason precisely about these ideas. The flavour of this branch of maths is often quite abstract (or even very abstract), because it explores mathematical questions for their intrinsic interest. However, the ideas and tools developed in this way have turned out to be crucial for many applications and have had their own profound consequences for the study of mathematics. The habits of mind developed in the course of studying pure mathematics are valuable in a range of careers.

Topics in a maths degree that might be described as pure mathematics include the following.

Abstract algebra (group theory, linear algebra, etc.). Abstraction is one of the concepts that makes maths so powerful. This starts very early on: for example, when we recognize that there's a notion of 'five-ness' (when we talk about five apples, or five cats, or five books, or five colours, the 'five' is the same in all cases). We can reason abstractly about numbers – say, $5 + 3 = 8$ – without needing to know whether we're considering apples or cats or books or colours. Abstract algebra takes this idea further. When we notice the same underlying algebraic structure, we can distil out this structure and then reason about all objects with the same structure. One important example is that of a vector space, which underpins linear algebra (see Chapter 5).

Another key algebraic object is the group, which consists of a set with an operation that satisfy certain properties. Examples of groups include the set of integers (whole numbers) under addition, the set of non-zero complex numbers under multiplication, the set of translations of the plane under composition, and the possible moves of a Rubik's Cube. Group theory is important in many areas of maths and has applications in crystallography, computer science and particle physics, to name just a few examples. There are other important algebraic structures beyond vector spaces and groups, which you might study in a more advanced abstract algebra course.

Analysis (real and complex). Imagine drawing a continuous function on the usual x–y axes that's negative for one x value and positive for another. The function must take the value 0 somewhere between these two x values: it must cross the horizontal x-axis. This seems intuitively clear, but how might we prove it? (The result is called the intermediate value theorem, by the way.) Before we can do so, we need a careful definition of continuity. A definition such as 'the function can be drawn without taking your pen off the paper' is good for intuition, but it's not sufficiently precise to prove a result like this.

A course on analysis will include formal definitions of concepts such as the convergence of a sequence or series, the continuity of a function, differentiability and integrability. From these definitions, a theory can be developed via a careful sequence of robust proofs. Such theories give the justification for many of the results in calculus that students learn at school or college, as well as the more advanced theory that students meet during a maths degree. Particular branches of analysis have their own focus: real analysis concentrates on functions from real numbers to real numbers, while complex analysis studies functions whose inputs and outputs are complex numbers. This turns out to be enormously useful in applications such as fluid dynamics, and it's intriguing to explore in its own right.

Combinatorics and graph theory (also called discrete maths). The London Underground map is a well-known example of a 'graph' or 'network' in the sense of graph theory: it has stations (which are called 'vertices' or 'nodes' in maths) that are connected by Tube lines (or 'edges'). This map doesn't accurately capture the geographical locations of the stations or the precise routes of the train lines, but it does capture the important information: namely, which stations lie on which lines, and where it's possible to change between lines. Graphs and networks are widely used to model and analyse a variety of real-world situations: connections on a social media network, transport infrastructure in a city, the structure of the brain, the National Grid (electricity network), the spread of a disease – the list goes on.

In pure mathematics, the focus is on the abstract theory, with many elegant and surprising results (see our section on Ramsey theory in Chapter 7). The broader theory of combinatorics includes versatile techniques for counting and other 'discrete' problems (as opposed to the 'continuous' problems studied in calculus). There are powerful combinatorial techniques using probability, and there are many connections between combinatorics and computer science, as well as links to other branches of mathematics and to statistical physics.

Geometry. Students start learning geometry at school, where they study Pythagoras's theorem about right-angled triangles, the circle theorems, and coordinate geometry (using equations of lines and circles). Geometry at degree level moves beyond this. You might study vectors in two and three (and more) dimensions. You might study other coordinate systems, such as cylindrical and spherical polar coordinates (recording points using a suitable combination of distances and angles). You might study curvature (how curved is a certain curve or surface?). You might learn about non-Euclidean geometry such as spherical geometry and hyperbolic geometry (see Chapter 7). You might study the interaction between geometry and symmetry (group theory). You might work on metric spaces, where the notion of 'distance' is generalized from the familiar Euclidean distance. In addition to the intrinsic interest of these geometric topics, many are relevant in other areas of maths. For example, a good understanding of vectors and coordinates in two and three dimensions is needed for multivariable calculus, which in turn is used for fluid dynamics, electromagnetism and more, and there are many links between advanced geometry and topics in theoretical physics such as general relativity.

Logic. Is every true mathematical theorem provable? Is it possible to prove a false theorem? These are some of the deep questions at the heart of maths. The concept of proof is central to mathematics: we start with some known assumptions ('axioms') and rigorously proceed, via a series of careful deductions, to a theorem. So what should these axioms be? There are different possible answers to this, but it turns out that each necessarily has its limitations. Logic explores these foundational questions about truth and provability, and it's closely connected with the philosophy of maths. There are also computer science aspects to it: for example, the concept of 'proof verification', where a computer will systematically check that every proposed deduction in a suggested proof is logically valid and fully justified.

Number theory. To describe number theory as the study of the properties of the integers (whole numbers) is to undersell it (although this description *does* have a certain amount of truth to it). Number theory explores the properties of prime numbers (more on this in Chapter 7) and divisibility, and it seeks to detect and explain structure within the integers.

One type of equation you might come across in number theory is a Diophantine equation: this simply means an equation where we seek whole number solutions. Solving such equations can be extraordinarily difficult (see Fermat's Last Theorem in Chapter 7). Advanced number theory includes the analytic branch, that is, using ideas of calculus/analysis in powerful ways, and the algebraic branch, that is, generalizing the whole numbers to broader classes of interesting numbers, such as rings and fields (this is closely related to abstract algebra; see above).

Although sometimes seen as one of the purest forms of maths, modern number theory has important applications in cryptography, and it even has links to physics via an apparent connection between the zeros of the Riemann zeta function and random matrix theory.

Topology. If we look at a surface, how can we tell how many holes it has? For example, a tennis ball has no holes (it's a sphere), whereas a ring doughnut has one hole (it's a torus). And when can we 'continuously deform' one surface into another? For example, if we have a clay sphere, can we squeeze and stretch it (but not tear it) in order to make it into a torus? Topology addresses questions such as these: problems where there are important notions of shape, but where distance is not important. Topology can be studied not only in three dimensions, but also in higher-dimensional space. One aspect of topology is knot theory, which explores questions about whether two knots are essentially the same; this turns out to have applications in the folding of DNA. Topological thinking is widespread in pure mathematics and has connections with both theoretical physics and data analysis, for example.

Applied mathematics

Applied mathematics is about solving the problems that arise in other disciplines using mathematical ideas and tools. There's a long history of mathematics being used in physics and astronomy, for example, and this continues today. Mathematicians also work alongside biologists, medical professionals, economists, chemists, engineers and social scientists, to give just a few examples. In the context of university courses, some applied mathematics modules explore these applications and introduce modelling techniques that mathematicians can use to turn messy real-world problems into problems that are able to be addressed using mathematical tools.

Other modules are 'methods' courses. These are about learning key tools and techniques that are useful in applications, with a focus on understanding how the techniques work rather than on providing a rigorous explanation of their mathematical foundations (which would fall under pure mathematics).

Some modules are about techniques that can be explored and applied using pen and paper, while others will use technology – spreadsheets, general programming languages or specialist mathematical software – to carry out calculations and implement algorithms that would be impractical to attempt by hand.

Topics in a maths degree that might be described as applied mathematics include the following.

Calculus. Calculus is one of the most fundamental ideas in mathematics, allowing us to make sense of questions about the rates of change that occur in many applications. There are two main aspects of calculus – differentiation and integration – and these are intimately connected. The concepts of differentiation and integration you've encountered at school or college are extended to higher dimensions at university: multivariable calculus, for instance, is important for lots of applications. Many of the topics in the list below rely on, and develop further, the ideas of calculus.

Differential equations (ordinary, partial and stochastic). A differential equation is an equation that involves one or more derivatives (rates of change). As with other types of equation, the goal is to solve the equation, where possible, and to use it to learn as much as we can about the context in question. Differential equations occur in all shapes and sizes. Those where the rate of change is always with respect to the same variable, such as time, are called ordinary differential equations. Meanwhile, partial differential equations allow for more variety, such as handling situations with more dimensions. If there is randomness involved, then stochastic differential equations come into play; these are mentioned again under the 'statistics' heading below. We'll meet ordinary and partial differential equations again in Part II of this book, reflecting their widespread importance in applied mathematics.

Dynamical systems. Lynx eat hares. There's a delicate relationship between the number of hares and the number of lynx in a given geographic area. If there are lots of hares, then there's plenty of food for the lynx, so the lynx population grows. However, when there are many lynx, they eat lots of hares, so the hare population drops. Food is then scarce for the lynx, so *their* numbers drop again, allowing the hare population to recover.

This relationship between numbers of hares and lynx can be modelled mathematically using a dynamical system known as predator–prey equations. A dynamical system is one that evolves over time according to a mathematical rule. The predator–prey equations are differential equations, capturing the mutual interdependence of the numbers of lynx and hares. It's not typically possible to find an exact solution for a dynamical system, so techniques instead focus on other ways to analyse the situation. For example, it's useful to illustrate the numbers of hares and lynx on a graph, with hares on one axis and lynx on the other, to show how the system evolves from various different starting points. There are many real-life instances of dynamical systems. A course on dynamical systems will include a range of mathematical techniques to explore the long-term behaviour of dynamical systems, which can sometimes be very subtle. The study of dynamical systems also includes the mathematical theory of chaos.

Dynamics and mechanics. Newton's law of gravitation allows us to study questions about orbiting bodies. When, for instance, will the International Space Station next pass over your house? What trajectory should the BepiColombo mission follow on its way to Mercury? The study of these questions draws on the ideas of differential equations and multivariable calculus mentioned earlier in this section. Although Newtonian mechanics has in some sense been superseded by more modern physics, its ideas are still hugely powerful, and the answers provided by this theory are sufficiently good for many questions. More modern formulations of these ideas, such as Lagrangian mechanics and Hamiltonian mechanics, link to other aspects of physics, such as the famous theorem of Emmy Noether. A good understanding of dynamics and mechanics is needed to access many other aspects of mathematical and theoretical physics.

Fluid dynamics. Perhaps unsurprisingly, fluid dynamics is about the motion of fluids (liquids and gases). It has close links with physics and engineering, and it makes heavy use of a range of mathematical tools. There are different models available, depending on the fluid in question and the level of sophistication required. In some contexts, such as the flow of air past an aerofoil, or the motion of water waves, Euler's equations will suffice, while in others, viscosity (a sort of internal friction) is important. Branches of fluid dynamics include aerodynamics (whether for an aeroplane in flight or for a cyclist in a velodrome) and magnetohydrodynamics (which has applications to the study of sunspots).

Fluid dynamics is used to study the weather and the climate as well as the formation and melting of sea ice and glaciers. It's also used to study biological applications, such as the flow of nutrients through a placenta. As with all applied mathematics, it involves selecting and refining models in light of real physical data. It's often helpful to use computational techniques to explore the physical system and the associated mathematical theory.

Mathematical modelling. This is a broad heading that encompasses the wide variety of ways in which mathematics is applied to make sense of the real world. When looking to model a real-life situation, mathematicians must identify the key factors that their model needs to capture, distilling these out from the surrounding complexity and making simplifying assumptions to render the situation tractable. They can then formulate a model, perhaps consisting of one or more equations that describe the relationships between the relevant quantities. The next step is to explore the model, by either solving the equations by hand or (more probably) using computational techniques to see how the situation evolves, or to get an approximate solution. The information gleaned can then be translated back to the real situation in order to make predictions.

Often mathematicians test their models by applying them to situations where the outcome is already known, to see whether the model is producing useful information. It may be necessary to refine the model if it has omitted factors that turned out to be important or is in other ways unrealistic.

If this all sounds very general, that's because it is: this philosophy is used whenever maths is applied. There's more discussion of this in Chapter 5.

Numerical analysis. Meteorologists use mathematical models to predict the future behaviour of the atmosphere given the current starting conditions. To create such a weather forecast, meteorologists have to solve these equations – that is, find a very close approximation to a solution – and this is hugely computationally demanding, even with the supercomputer the Met Office uses to perform the calculations. Numerical analysis provides the techniques and understanding to be able to do this. Its focus is on algorithms for finding solutions, or approximate solutions, to computationally demanding problems in a sensible amount of time and with a high level of reliability. These problems might, for example, involve manipulating large matrices, as is discussed in Chapters 5 and 6, or solving large systems of interconnected partial differential equations. Numerical analysis is important everywhere computers are used to assist with scientific calculations: processing data from medical imaging, designing jet engines and, in fact, just about every application of maths mentioned in this book.

Optimization. Optimization is all about trying to 'optimize' a quantity, subject to some constraints. This is a rather general description, reflecting the fact that this topic has many facets and is applicable in many ways. For example, say a delivery company has a warehouse full of parcels to deliver, and some delivery drivers and vans available. How should they divide up the deliveries into routes? In this case, the quantity to be maximized is the profit, with the constraints of where the parcels need to be delivered, the numbers of drivers and of vehicles, and the lengths of the drivers' shifts. Sometimes optimization problems can be solved exactly (as in, there's a 'best' solution and it can be found in a reasonable amount of time). Sometimes this is impossible, or at least impractical, so the focus is on finding a good enough solution in the available time: a solution that's known to be within at least 5% of the best possible answer and can be obtained in a few minutes might be much more useful to a business than an answer that's the best but takes weeks to find. There are different branches of optimization, which tackle different types of problem and use different techniques. There's more about one branch, linear programming, under the 'operational research' heading below.

Quantum mechanics. You'll find courses in quantum mechanics in both maths and physics degrees. Quantum mechanics is central to modern approaches to understanding the natural world at the atomic scale. One key idea is the Schrödinger equation, which is a differential equation that describes the wave function of a particle such as an electron. Solving this equation allows us to study the behaviour of such particles.

Quantum mechanics is a theory from physics that makes verifiable predictions about the behaviour of the real world: for instance, Albert Einstein's Nobel Prize was for his work on the photoelectric effect, which is part of quantum theory. It's also a rich mathematical theory, drawing on ideas from probability, linear algebra (eigenvalues and eigenvectors), functional analysis, representation theory and more.

Relativity (special and general). As with quantum mechanics, you'll find special relativity and general relativity in physics degrees as well as in maths degrees. Special relativity allows us to combine the notions of space and time so that we can work with space–time. This is relevant when we're studying objects moving very fast: specifically, moving at speeds close to the speed of light. This isn't just theoretical, either: particle accelerators exist to accelerate particles to travel at close to the speed of light.

One principle of special relativity is that the speed of light is a constant. This takes some unravelling. If a friend runs towards you while you're standing still, they'll appear to move slower than if they were to run at the same speed (from their perspective) while you were also running towards them: the speed of your friend (from your perspective) also depends on your own speed. But light is somehow different. To make sense of the idea that the speed of light is constant, special relativity introduces the idea of time dilation, famously illustrated by an experiment in which one atomic clock remained stationary on earth while another was sent around the world on a very fast aeroplane. As predicted by special relativity, the clock that travelled ran slower than the clock that stayed at home. This phenomenon has practical significance: for instance, with satellite signals being transmitted for GPS navigation.

General relativity goes even further than special relativity, providing an explanation for gravitation and addressing concepts such as black holes. Like special relativity, general relativity provides theories that can be (and have been) tested experimentally, and it relies heavily on mathematical ideas. For example, special relativity uses Minkowski space to capture the notion of space-time, and this is different from familiar Euclidean space. General relativity takes this even further, connecting to ideas of advanced geometry, for example.

You'll also find maths applied to specific topics in biology, chemistry, climate science, economics, engineering, finance, medicine, physics and more.

Statistics

Statistics is about making sense of data and understanding uncertainty. It's closely interconnected with probability (indeed, probability provides the theoretical underpinnings for statistics) and has never been more important than it is in the twenty-first century. Modern computing techniques allow companies, governments and researchers to generate huge volumes of data, which then need to be analysed in order to provide useful information. What conclusions can we draw from a set of data? How confident can we be that these conclusions are correct? Addressing these questions uses tools from mathematics and beyond.

Topics in a maths degree that might be described as statistics (or probability) include the following.

Actuarial science. Actuarial science is about quantifying risk so that it can be managed appropriately. This occurs in a range of contexts, many – but not all – of which are financial. For example, what premium should a company selling car insurance charge each customer? Actuarial science is important in the field of pensions as well as insurance. Let's say someone who's about to retire has accrued a pension fund and wishes to purchase an annuity (where they pay their pension fund to a company and, in return, receive a regular payment for the rest of their life). How much should the company providing the annuity charge up front for a given level of annuity? Another potential use of actuarial science is in investment. How risky are the assets (government and company bonds, company shares, etc.) being considered by an investment fund? Actuarial science develops mathematical models to address questions such as these.

Data analysis. Modern computing techniques allow us to generate and manipulate much larger quantities of data than was possible even a few decades ago. The Human Genome Project, for instance, generated massive amounts of data about human DNA. A supermarket loyalty card scheme also generates huge volumes of data about customers and their shopping habits. In order for any of this data to be useful, it's not enough to simply collect it and store it: we need to analyse it. What are the key patterns revealed by the data? What conclusions can be drawn? How confident can we be in the validity of those conclusions? It's also important to be able to present the data in a way that clearly and accurately conveys the key points. There's more about this in Chapter 6.

Machine learning. One approach to getting a computer to analyse data is to tell it exactly what to look for, and how. Machine learning uses a different strategy: the computer teaches itself what patterns to recognize. This turns out to be potentially very powerful in many applications, such as in medical diagnostics. A computer can be shown lots of data from a certain type of scan and told which correspond to positive or negative diagnoses for a particular condition. The computer deduces what patterns and structures to look for in the data, and it can then go on to make its own diagnoses (of scans that have not first been assessed by a clinician). Another use of machine learning is in recommendation systems, such as when YouTube recommends what video you should watch next. Behind the scenes, the software gathers lots of data on which videos are popular with which viewers. Rather than being programmed in advance to know what the key factors are, the system can interrogate this data to learn viewing patterns and then use that information to recommend further videos. Machine learning draws together techniques from maths, statistics and computer science.

Markov chains. Markov chains build on other ideas from probability (see the separate 'probability' section below). A Markov chain is a way to model a certain type of process, consisting of a sequence of states according to some probability distribution, where each state depends only on the previous one. A common example in introductions to Markov chains is the 'gambler's ruin' problem. Imagine a gambler starts with £100. Once a minute, they bet £1 on the outcome of a coin flip: they win £1 if the coin comes up heads, and they lose £1 if the coin comes up tails. The coin used could be a biased coin, where heads and tails are not equally likely, but the chance of each is known. What do we expect to happen in the long term? Will the player eventually lose all their money, and, if so, when do we expect this to happen? There are many ways to analyse this problem, but Markov chains give us the following approach: at any stage, the amount of money the player has depends only on the amount at the previous stage and on the outcome of the coin flip. In other words, this amount isn't affected by the earlier sequence of flips. The advantage of the Markov chain approach is that it generalizes to other, more complex problems, and Markov chains (and, more generally, Markov processes) have many applications. For example, some speech recognition systems are built on the theory of Markov processes.

Monte Carlo simulation. Imagine you want to estimate the area of Manhattan Island. One strategy to do this is as follows. Take a rectangular map of Manhattan Island. Choose a uniformly random point on this map and record whether or not it lies inside the island. Do this over and over again, each time choosing a uniformly random point on the map and recording whether or not it lies inside the island. If you do this with enough uniformly random points, then you'll get a good estimate for the area of Manhattan Island: the proportion of points lying inside the island, multiplied by the area corresponding to the map. This is an example of a Monte Carlo method, using randomly chosen points to estimate a quantity that it would be impractical to work out exactly using a deterministic (non-random) process. Monte Carlo simulation has many varied applications. For example, if you're planning a new bus timetable, you'll want to make sure that there are enough buses at peak times, but not lots of nearly empty buses driving around at quiet times. Monte Carlo simulation techniques can be used to give a computer-based model showing what might happen with different timetables, with passengers arriving at the bus stop at random times but with frequencies informed by real data. This relates to the broader category of simulation in the operational research topics we'll discuss shortly.

Probability. It's not quite right to put probability under the heading of statistics, but the two subjects are certainly connected. Probability gives us the vocabulary and techniques to study questions about chance and risk. It's a fascinating subject in its own right, with connections via measure theory to analysis. It's also fundamental to the study of statistics, giving us theory and tools to make sense of risk and uncertainty. So a grasp of probability is needed for many of the topics under this statistics heading. The study of probability often begins with questions about coin flips and dice, but it develops a lot from there. For example, Markov chains (see above) is a branch of probability. More advanced topics in probability include stochastic processes, such as Brownian motion, and stochastic differential equations, which can be used to model situations where randomness is involved.

Statistical inference. In the run-up to the 2016 Brexit referendum in the UK, polling organizations ran opinion polls to try to gauge what the result would be. It wouldn't be practical to ask the whole electorate of tens of millions of people how they intended to vote, so instead the pollsters conducted surveys of hundreds or thousands of potential voters. They then needed to infer from this sample data what the overall view of the population was at the time. This is a good example of a situation in which statistical inference is necessary. Statistical inference can be used to estimate a parameter (such as the proportion of the electorate who would vote in a particular way if the referendum were held that day) as well as to estimate confidence intervals. For example, when the Office for National Statistics (ONS) reports an estimate of how many people in the UK are unemployed, it does so using the Labour Force Survey, so its figure is an estimate. The ONS also publishes the 95% confidence interval, which is a range in which it's 95% confident that the true figure lies. This gives a more nuanced understanding of the data. There are different approaches to statistical inference, such as the frequentist approach and the Bayesian approach, which reflect different perspectives on how best to draw robust conclusions from the data.

Applications of statistics to particular areas. Statistical ideas, tools and techniques are applicable in any context involving data analysis or understanding uncertainty. The topics above are widely applicable, and many university modules are designed to help students to learn these general techniques. There are also modules that focus on specific applications, exploring the relevance of standard statistical tools and developing tailored theories relevant to a particular context. For example, a course on financial mathematics might include the Black–Scholes formula, which is a partial differential equation describing the price of a financial option.

Statistical models and techniques are also important in genetics, for example, in seeking to understand key mutations and in identifying which gene or genes control a particular attribute. A course on medical statistics might include the subject of experimental design. If researchers wish to test the effectiveness of several different treatments on a group of patients, then they'll need to consider how to allocate those interventions in such a way that the resulting data is useful. If it turns out that one intervention was used exclusively on patients who had a less severe version of the condition while another intervention was used exclusively on patients with the most severe version, then it will be problematic to infer a robust conclusion from the result.

These more specialized courses on particular applications of statistics tend to be offered in the later years of a degree programme, once students have got a secure grasp of the fundamentals of probability and statistics in their earlier years.

Operational research

Operational research, or OR, is about using mathematical and statistical techniques to analyse complex real-world situations to inform better decision making. OR grew from work carried out just before and during World War II, when mathematical tools and approaches were deployed to make more efficient use of scarce resources. In the twenty-first century, OR is used widely by businesses, governments and charities, and it's sometimes called 'management science'. Professionals working in OR often have the word 'analyst' in their job title.

Topics in a maths degree that might be described as OR include the following.

Algorithms. An algorithm is a precise set of instructions defining how to complete a certain task. In order to be useful, an algorithm will need to terminate (finish) after a finite length of time, and this time needs to be short enough to suit the particular context. When designing or studying an algorithm, one important question to ask is how long it will take to run for a particular size of input; this is measured as the complexity of the algorithm. The design and analysis of algorithms draws together ideas from maths and computer science. We'll revisit the importance of efficient algorithms in Chapter 6, where we discuss their use in a UK kidney donation scheme.

Forecasting. Businesses need to be able to predict future demand. If you run a retailer, then you might need to order stock for Christmas months in advance. In that case, you'll need to know what to order and in what quantities to meet customers' demand, without having lots of surplus. If you run a hospital, then you might be required to predict what future patients will need, so that you can ensure resources (staff, beds, operating theatres, and so on) are allocated appropriately. There are sophisticated techniques available to help with these sorts of decisions, and they come under the heading of forecasting.

Game theory. The word 'game' here might trivialize a subject with far-reaching implications. The mathematician John Nash (whose life inspired the film *A Beautiful Mind*) is just one of the people who have won the Nobel Memorial Prize in Economic Sciences for their work on this subject. Game theory gives precise mathematical tools to analyse situations involving two or more 'players', where each has to make a decision about how to behave, and where these decisions have an effect on the other players. We might want to answer the question 'What's the best strategy?', or even 'What would it mean for a strategy to be "best"?'. A simple example of a situation where this theory could be used is the game 'rock, paper, scissors': each of two players chooses one of the three outcomes, and for each possible pair of outcomes, either one player wins and the other loses, or it's a draw. A game such as this can be analysed mathematically to try to determine the best strategy. At a more profound level, similar techniques from game theory can be used to study human behaviour, for example, in politics, economics, business, retail, biology and beyond.

Linear programming. Imagine a business that can make two products, A and B, in a particular factory. Product A is time-consuming to make but sells at a high profit. Product B is quick and cheap to make but has a low profit margin. The company wants to maximize its profit, with a given amount of factory space and staffing level. Should it focus entirely on Product A, or entirely on Product B, or produce some combination of both? More sophisticated versions of this problem would be more realistic, perhaps taking into account the demand from customers for each product, the cost of storing products until there's a demand for them, the cost of changing production lines from manufacturing one product to another, and so on. Linear programming is used to find optimal solutions to problems of this type. This is a branch of optimization, which was included under the 'applied mathematics' heading above – linear programming and optimization could be considered as both applied mathematics and OR.

Simulation. Making any sort of change in an organization can have significant consequences, whether intentional or not. Simulation is a valuable tool to help us predict what the consequences of a change might be. In the same way that an architect might create a physical model of a building they've designed in order to help their client see what's being proposed and what the knock-on effects might be, operational researchers can create a computer-based model of a business and simulate the effect of proposed changes. Choosing an appropriate model and interpreting the results carefully requires particular OR expertise.

But maths doesn't really fit into neat subheadings

We've just met a lot of topics; don't worry if you feel a bit overwhelmed. I hope the above has given you a glimpse of the variety of subjects that might be included in a maths degree. Importantly, there's no clear dividing line between these approximate subdivisions. For example, a 'pure' approach to graph theory or network theory will look at abstract graphs and prove theorems that apply to all graphs with certain properties, while an 'applied' approach might focus on a particular graph or network arising in practice and seek to apply mathematical tools in order to understand its behaviour. Similarly, mathematicians approach a topic such as partial differential equations from different perspectives: they might seek rigorous proofs of certain results and identify the precise conditions under which those results hold, or they might adapt numerical techniques for solving these equations in order to model a real-world situation. Another example is probability, which is important in statistics but can also be studied from the perspective of pure mathematics.

There are also topics that will sometimes be offered as part of a maths degree that don't fit neatly into the above categories. These include maths education, maths outreach and communication, and the history of maths.

Some maths degrees include a module about careers and employ-ability, sometimes concentrating entirely on developing skills relevant for future employment and exploring the options open to maths graduates, sometimes addressing these points alongside some aspect of maths. Other degrees don't explicitly have modules about careers and employability, but students are able to develop their knowledge and skills in these areas in other ways. We'll see more on this in later chapters.

In a similar vein, many maths degrees include some aspect of com-putational maths, perhaps as the main focus of a module or perhaps as a skill developed while learning some new mathematics. This might involve learning to code using a general programming language such as Python, or getting to grips with specialist mathematical or statisti-cal software such as Mathematica, MATLAB, Octave or R.

The key point is not really which language or software you learn, because once you've got one of these under your belt, you'll find that you can pick up others quite quickly. Rather, it's to learn the general principles (which, while general, are best learned in a specific context): how to structure an algorithm that a computer can implement, what a computer can do for you to help with solving a mathematical or statistical problem, how to interpret (but not over-interpret) the output from a computer, and so on. You might be pleasantly surprised to discover how sophisticated these tools are, meaning that your computer can focus on routine, repetitive, potentially lengthy computations while you turn your attention to problem solving, modelling and interpretation.

Moving from school maths to university maths

As we've seen, some topics in a maths degree build on those you've already studied, whereas some will be new to you. There's a good chance that you'll revisit some familiar content at the start of your degree. One reason for this is that different people are familiar with different topics (because they've studied different qualifications at school, perhaps), and universities need to ensure that everyone is comfortable with some core material before going any further. But even if you *have* met certain topics before, they might be taught in a style that's very different from what you've experienced at school or college. You might move through the material faster now that you're at university. You might explore familiar subjects in more depth, seeking to make connections between topics and placing more emphasis on understanding why certain procedures and techniques work, in addition to knowing how to apply them. This can help you to get a fresh perspective on maths you thought you already knew, and reaching a deeper level of understanding can be very satisfying. It's a bit like having a car: for some people, being able to drive safely from A to B is enough, while for others, looking under the bonnet and knowing how the engine works is important too.

Sometimes students are surprised to find that a topic described to them as 'pure' at school or college is considered as 'applied' at university. For example, the techniques of differentiation and integration are often considered to be pure mathematics at school or college, but they're typically studied in applied methods courses at university. You might meet a new topic called 'analysis', which is a rigorous 'pure mathematics' take on calculus. If you study analysis, you'll explore a precise approach to differentiation and integration, using careful definitions and proofs to ensure the subject is built on logically firm foundations.

On the other hand, statistics and probability are often described as 'applied' at school or college, but at university, students begin to explore the many connections between these topics and ideas in pure mathematics.

You might have heard people talking about there being a jump in difficulty from school maths to university maths. While there are certainly differences between the two, this isn't the same as the latter representing a massive step up in difficulty. Some of the differences might be to do with teaching approaches, with the style of questions you're being asked to tackle, or with the amount of independent study you're being expected to undertake. Others might be due to the fact that you're meeting new topics.

It's important to remember that universities know the move from secondary to higher education involves a transition to new material and new ways of working: degree programmes are designed to support students making this transition. Universities are familiar with the profiles of their incoming students and what they can expect them to know or to be able to do, and they tailor their programmes accordingly. Having said that, you'll still need to adapt to new ways of working, to be willing to persevere and to ask for support when you need it. There's no need to panic, though. Over the course of a three- or four-year maths degree, you'll become increasingly independent and develop your study skills and mathematical sophistication. You don't need to have done all this before the first day of your degree!

The choice is yours

As we've discussed, often students leaving school or college haven't experienced the full range of topics that maths has to offer, so they don't know what they want to specialize in straight away. What's more, the bricks of the mathematical wall overlap each other; most areas of maths are closely interconnected, so – in order to reach the higher levels of this wall – it's important to have a solid and broad foundation. This is one of the reasons why many maths degrees include topics from all areas in the first year: it allows students to develop that broad foundation and to make informed decisions about what they'd like to study further. Even if you already know that you're interested in a specific application of maths, you'll probably need to take some modules that fall under the 'pure mathematics' heading in order to learn about and understand the tools and techniques you'll be applying later on. Similarly, even if your focus is pure mathematics, it's always helpful to study applied mathematics and statistics, as the ideas you encounter in these courses can be relevant for pure mathematics too.

In this chapter, we've seen an overview of the kinds of topic that are waiting to be explored at university. However, deciding which subjects you'd like to pursue is just one of the many factors you'll need to consider when choosing a course of study. For instance, there are a variety of ways these subjects can be combined to form a degree programme that suits your interests. And there are many other factors to consider, such as how teaching and assessment are structured, and what opportunities you'll have to interact with staff and your fellow students. We'll discuss these in more detail in Chapter 2.

CHAPTER 2

What are my options for study?

DEGREES IN THE MATHEMATICAL SCIENCES vary enormously – which is a good thing – so it's really worth doing some research to find courses that are a good fit for you. Unlike at school, at university there's no national curriculum, so even two programmes that are both called 'mathematics' can be very different – and that's before you start to consider the possibilities within the mathematical sciences and the options for joint degrees! This chapter offers a description of the different forms your studies in maths might take and suggests some factors you might want to consider when choosing the right maths degree for you.

What type of degree?

The main focus of this book is undergraduate degrees, which are also called first degrees or bachelor's degrees. You'll find more about postgraduate study in the following sections, but for now let's stick with undergraduate study. There's a huge range of undergraduate degrees that involve maths, all of which can lead to further study or to an exciting career. Some are concentrated entirely within the mathematical sciences and might be called Mathematics, Statistics, Applied Mathematics or Mathematics and Statistics, to give just a few examples. (When you're searching for courses, don't forget that 'mathematics' might not appear in the title.) Others incorporate subjects that aren't related to maths, or that work in partnership with maths to lead to an additional qualification (such as qualified teacher status).

An undergraduate degree will lead to a qualification such as a BSc (Bachelor of Science) or a BA (Bachelor of Arts). It doesn't matter much whether it's a BSc or a BA: the distinction is usually a historical hangover, rather than a statement about whether maths is a science or an art. Both qualifications are equally respected by employers and universities. Depending on the university and programme you select, it might be possible to continue for an additional year beyond the BSc/BA to get a Master's qualification, such as an MMath degree. There's more about this later on.

Joint degrees and specialisms

There are many, many joint degrees allowing students to combine maths with another discipline. Common subjects studied alongside maths are business, computer science, economics, education, finance, philosophy, physics and a range of languages, but it's possible to study maths with many other subjects too: history, music, sport – the list goes on.

The structure of joint degrees varies considerably from institution to institution. In some cases, you'll study two subjects with equal weighting, while in others you'll study a main subject and a subsidiary one. There might be a lot of flexibility (at least in your later years of study) regarding how much time you spend on each of the two subjects, or you might be more constrained.

Even if your degree course is simply called Mathematics, it may still be possible for you to take modules from other subjects. Many universities use a credits system, which requires you to take modules worth a certain number of credits in order to gain your desired qualification. Depending on where and what you study, you might be able to earn some credits from courses offered by non-maths departments. You'll need to explore the individual programmes at specific universities to find out what's permitted and where.

Of course, you don't always have to look outside of maths to add a little variety to your degree. A maths department might also offer modules such as the history of mathematics, maths education or maths outreach.

You should also research how much scope you'll have to specialize, particularly later on in your course of study, if that's important to you. Some universities, particularly those with large maths departments, offer a lot of options from which students can choose a few to study. These universities can do this because they have lots of students (so there are viable numbers of students for each option) and plenty of faculty with differing expertise to teach the options. This allows students to pursue their particular interests and specialisms. Other universities have a greater proportion of compulsory core modules and offer shorter lists of options from which students can choose their remaining modules. While this will give you less choice, if the subjects you're interested in are still covered, then the lack of choice might not be a problem. There might be other reasons you prefer these programmes too, such as the teaching and assessment methods.

If you feel that you already know where your interests lie or what your future career path will be, then you might want to choose a specialized degree programme to reflect this. However, it's worth considering more general programmes, as they might allow you to specialize in your preferred direction while also giving you greater flexibility if your interests or plans evolve. You'll find that many universities offer a suite of degree programmes in the mathematical sciences, and often they share the first year (or the first year overlaps significantly between programmes). This means that it's potentially easy to switch between programmes if, after a few months, you find that you're not in the right one for you.

The length of the course

How long is an undergraduate degree? The most common answer is three years, but it's a bit more complicated than that, as we'll see. Even among full-time degrees, there's some variation in the lengths of courses. Here are some reasons why a programme might not be a three-year programme leading to a BSc/BA.

Studying part time

If you study part time, then it'll of course take you longer than three years to achieve your qualification. Some students need to fit their studies alongside work, caring responsibilities or other commitments, and a part-time degree can be a good way to achieve this. There are various part-time options available to students looking to explore degree-level maths, some of which lead to degrees such as the BSc/BA described above, and some of which lead to other qualifications such as a Certificate of Higher Education. If you think that part-time study might be for you, you'll find plenty of useful information on the UCAS website as well as on individual universities' pages.

Differences around the UK

Scotland has its own system, which means that degrees often take a year longer to complete there than elsewhere in the UK. If you have the appropriate qualifications from school or college, then you may be able to start directly in the second year.

Foundation years

Some universities offer a foundation year in maths, which allows students to prepare themselves more fully for a degree. This means that it'll typically take four years to graduate with a BSc or BA. The foundation year helps students to get used to life and learning at university, to deepen their subject knowledge and to build their confidence before embarking on undergraduate study. A programme including a foundation year can be a good option for people whose educational profile doesn't meet a standard university offer, or who have been out of education for a while. You'll find these programmes on the UCAS website alongside regular undergraduate courses.

Work placements and studying abroad

Some programmes include a placement year working in industry. This can be a great way to explore the applicability of the maths you've been studying and to boost your employability. It will typically extend the length of your degree by a year.

Depending on the university and the programme, you might also have the option to spend a year studying abroad. In some cases this will add to the length of your degree, while in others this year will simply replace one that would otherwise have been spent studying at your home university.

A placement year, or year abroad, might be particularly import-ant to you, but there are lots of other ways you can incorporate real-world experience into your time studying. For instance, you might want to spend your vacations (particularly the long summer ones) travelling or obtaining work experience through a paid internship, or you might have the opportunity to undertake a work placement (not as long as a year) as part of your degree programme. Many universities offer support – for example, through a careers service – to help students find such paid internships.

Further study

Some universities offer an integrated four-year degree (or five-year degree, with a placement year) leading to a Master's qualification such as an MMath (Master of Mathematics) or an MSci (Master of Science). Studying for such a degree allows you to take your mathematical studies further, and it can be helpful when applying for jobs or further study.

When you apply to university, you'll need to specify which course you'd like to take. There are different UCAS course codes for each degree. In practice, however, BSc/BA courses and MMath/MSci courses often overlap a lot (or entirely) for the first year or two, so it's likely you'll have some flexibility and be able to switch between the two once you've started, if necessary. Keep in mind that you may also need to obtain high enough grades in your first three years at university to be allowed to continue to the Master's year of a specific course.

While you don't need to decide in advance whether you'll want to do the extra year for the Master's qualification, it's still worth considering. If you think that you might want to do a course that can include a Master's year, this could affect your choice of university. It should be noted, though, that if you complete a BSc/BA degree, you can still apply for a one-year Master's degree, either at the same university or at a different one. So even if you choose a university that doesn't offer an integrated four-year Master's degree, you might still have the option of continuing on to a Master's qualification.

There's a huge variety of Master's courses in the mathematical sciences: some specialize in a particular area, while others are more broad; some are taught, while others are more research based. Graduates wishing to conduct further study in maths might pursue a Master's course (such as an MSc) specializing in a particular area of maths or in a related discipline such as computer science or engineering. They might also opt for a research degree, such as a PhD, or a postgraduate course, such as a PGCE: a qualification for those wishing to become schoolteachers.

Of the people who continue on to further study, some will stay working in the university sector, perhaps teaching undergraduates or carrying out research, while many will move into other careers, including jobs for which a Master's or a PhD in a highly mathematical subject is necessary. Having a PhD in maths is a requirement for certain roles (such as that of quantitative analyst – colloquially known as 'quants' – in the finance sector, or that of modeller in a variety of industries). Opportunities for people with maths PhDs may include positions in universities, but they aren't restricted to them: businesses and governments also want to employ maths postgraduates for their advanced research skills and expertise.

Accreditation

In some careers, it's helpful, or even a requirement, to have additional specialist professional qualifications.

Some degrees, or modules within degrees, have particular accreditation. This means that they've been checked and approved by the named organization, so a student who has successfully completed this degree/module has automatically met the requirements of a professional qualification. Studying an accredited degree/module is not the only way to meet these requirements: you might be able to apply after graduating for your qualification to be recognized, or you might be able to take additional exams to meet the requirements. If the degree/module is accredited, however, then these steps can be waived, and by completing the degree/module, you've already taken certain steps towards the professional certification.

Within the mathematical sciences, there are a few relevant types of accreditation, which we'll look at in this section.

Chartered Mathematician

The Institute of Mathematics and its Applications (IMA) is responsible for awarding Chartered Mathematician status to people with an appropriate level of mathematical education and professional expertise.[3] These people can include 'CMath' after their name, showing their status as a professional mathematician. CMath status is perhaps less well known than the analogues in accountancy, architecture, engineering and other professions, but it's valued and respected just as much.

A Chartered Mathematician must have met certain educational requirements, and there are a few routes to this is: one is to have completed an MMath honours degree that has been accredited by the IMA for this purpose. Another is to undertake an accredited BSc honours degree together with further study or training at an appropriate level. Graduates with maths degrees that aren't accredited can still apply for Chartered Mathematician status, but they must demonstrate that they've reached the appropriate level of mathematical education.

Anyone wishing to apply to become a Chartered Mathematician must also have appropriate postgraduate training and experience, and they'll need to demonstrate that they're working as professional mathematicians. The educational requirements are only part of the process.

Chartered Statistician

There's an analogue of Chartered Mathematician status for statisticians, unsurprisingly called Chartered Statistician (CStat). This is administered by the Royal Statistical Society (RSS).[4]

The RSS runs an accreditation scheme related to both Chartered Statistician status and Graduate Statistician status.[5] A university can choose to apply to have a degree programme accredited, which involves the RSS confirming that a graduate of this programme has the appropriate level of technical skills and subject knowledge to be a statistician.

Actuarial science

If you want to be an actuary (a professional who analyses and manages risk), then you'll need to pass exams administered by the Institute and Faculty of Actuaries as well as to demonstrate an appropriate level of professional experience and development. Some university maths courses have been accredited for this purpose, which means that students who do sufficiently well on these courses receive exemption from some of the actuarial exams.

If you want to know more about becoming an actuary, then you'll find plenty of advice on choosing a degree programme on the website of the Institute and Faculty of Actuaries (https://www.actuaries.org.uk/).

Teaching in schools
Some degrees combining maths and education lead to qualified teacher status (QTS), which is a requirement if you want to teach in many English schools. Not all degrees in maths and education lead to this qualification straight away: many maths graduates who go into teaching complete a degree first, and then either take a one-year course called a PGCE or follow another route to obtain QTS.

So much for the different types of degree on offer. Now we're going to look at the basic requirements for getting into university to study maths, and what you can expect in terms of teaching and assessment, class and department size, and academic maths support.

Admissions requirements

What's known as a 'standard offer' will vary significantly between degree courses. Some universities frame their standard offers in terms of UCAS tariff points, while others specify particular grades in specific qualifications. It's important to find both a standard offer and a course that's a good fit for you. Note that 'a good fit for you' doesn't have to mean picking a course with a standard offer that's around the level of your predicted grades. Some students find that the course most suitable for their particular interests and preferences has a lower standard offer than they expect to achieve; that's no reason not to choose such a course.

If you're studying A levels, you'll find that many courses in maths will require a certain level of attainment in A level Mathematics (or equivalent). Some require, or strongly recommend, students to have taken Further Mathematics too, perhaps to AS level or perhaps to full A level. The Advanced Mathematics Support Programme (see the 'further reading' section at the end of this book) supports schools and colleges to offer Further Mathematics, and it has free resources available to help students.

Many university applicants are studying, or have previously studied, for other qualifications, and if this applies to you, then you'll need to research individual courses to find out whether you meet the requirements. If you're returning to study as a mature student, it might be worth contacting individual universities to discuss the suitability of your qualifications and other relevant experience.

Generally speaking, it's helpful to do as much mathematical preparation as you can at school or college, as this will help you get off to a flying start in your degree.

You may need to meet other requirements too, such as GCSE attainment (or equivalent). If English is not your first language, then you might be asked to demonstrate that your English language skills are at a good enough level to allow you to engage fully with the degree course. Again, it's best to check individual university requirements to see what they're looking for. Even for maths, it's important that your English (both written and spoken) is strong enough that you can make the most of the teaching you receive and feel comfortable expressing your ideas.

Additional tests

A handful of universities have additional tests relating to maths. At the time of writing, these include TMUA (Test of Mathematics for University Admission), MAT (Mathematics Admissions Test) and STEP (Sixth Term Examination Papers). Some of these exams are sat in the autumn before offers are made, and some universities use the results either to help them decide whether to make an offer or to inform the level of offer they make. Others are sat in the summer, and some universities make conditional offers that include a requirement to achieve at a certain level in these extra qualifications. It's definitely worth researching them all at an early stage in order to get to grips with the process.

You'll find past papers and other preparation materials freely available online, so you can do some preparation to get used to the style of questions asked in the relevant test. This style might be quite different from what you've experienced at school or college, which can make the questions look difficult at first glance. The more you practise using past papers, however, the more familiar with this style you'll become, and the more confident you'll be about tackling them when the time comes. This type of preparation is useful not only for these tests, but also for studying maths at university in general: practising solving maths problems is always time well spent!

Teaching and assessment

When you're exploring different courses, it's a good idea to investigate both how you'll be taught and how you'll be assessed. You'll find a variety of programme structures, teaching methods and assessment models across different maths degrees. You should consider what you think will work for you, given your current strengths, while also taking into account areas of potential development for the future.

Studying at university

You're likely to find that your time at university is organized quite differently from what you've experienced at school or college. In fact, one of the main differences you'll notice is that much of your time won't be organized for you at all: you'll have a lot more time for independent study. In addition, you'll experience new styles of teaching, and you'll develop effective working patterns and independence that will be invaluable to you both at university and beyond.

Employers value the skills that new maths graduates can bring to the workplace, and the various modes of teaching and assessment offered at university are very useful for developing these. For example, having the opportunity to work on an extended project, dissertation or thesis can be a great way to practise your research skills, to work independently and to develop the ability to communicate complex ideas effectively. Even if you're someone who prefers working on your own, having to undertake group tasks during your degree will help you to gain attributes that will be incredibly useful in your future career.

Lectures

Traditionally, a lot of university maths teaching has been delivered via lectures, to audiences ranging from 20 students to 200 or more. These days, lectures are often accompanied by online notes and may be recorded so that students can watch them back afterwards. Don't be surprised by 'old-fashioned' maths lectures, with a lecturer writing on a large area of whiteboards (or even blackboards) rather than using an interactive whiteboard or slides. While slides or computer animations can be ideal for communicating a particular mathematical concept, sometimes, when you're working through a calculation or a piece of reasoning, the old technology is still best. Writing by hand on a whiteboard helps the lecturer to talk through the thinking behind their argument, and it allows students to see the whole piece of reasoning (which may be lengthy!), not just the most recent step.

Lecturers are increasingly exploring other models of teaching, however, such as 'flipped classrooms': this is where students do some assigned reading ahead of the lecture, and then class time is spent discussing the particularly challenging aspects of the mathematical ideas or working together on related problems.

Classes/tutorials

You'll probably find that you spend some of your time in classes or tutorials. These words can mean different things in different universities, with a class or tutorial consisting of anything from a group of two or three students with one teacher through to a group of 30 students, depending on where you're studying.

Comparisons of contact time between universities sometimes overlook differences in class size: an hour working with a couple of students and a lecturer or graduate student is a very different learning experience from an hour spent in a class of 30 students.

The ways in which classes and tutorials fit into the bigger picture vary too. In some programmes, you'll use class time to work with fellow students on problems, with a lecturer on hand to discuss them either in small groups or with the group as a whole, much as you might have experienced at school or college. In other programmes, you'll be expected to have tackled some problems independently, and then the class or tutorial will focus on discussing these problems, sharing ideas and addressing particularly difficult concepts or techniques.

Independent study

As we mentioned earlier, you're likely to spend a good chunk of time working independently on problems sheets and projects. You might do this completely by yourself, or be encouraged to collaborate with other students, or be required to do some work in a group.

You might like to explore how much 'formative feedback' you'll receive on your work and what form it will take. This type of feedback (as distinct from 'summative' assessed work contributing to your degree) is designed to help you identify the strengths of your work and the areas for improvement, as well as to address any misconceptions or confusion you have about the content. You might receive all your feedback through discussions with your lecturer as they come round to see how you're getting on during a class, or you might be asked to submit written work from your independent study time to be marked, or you might be expected to complete online tasks where you get instant feedback. Practices vary significantly between institutions. In some universities, your feedback on written work will focus on assessments that contribute towards your degree (there's more on this later) and might take a few weeks to be returned to you, so the majority of feedback you receive will be from discussions in class. In others, you'll receive feedback on written work every week, within a couple of days, on top of feedback from classes.

Maths departments are increasingly starting to use online assessment tools to provide instant feedback on regular homework, and potentially also on examined work counting towards the degree. The best of these e-assessment packages are mathematically sophisticated, based on a computer algebra system that can check a student's working as well as a final answer, and that, where appropriate, can mark as correct any answer that's mathematically equivalent to the version stored by the software. You might be able to use these tools quite flexibly as part of your independent study, allowing you to do more practice on topics where you're less confident, while quickly checking your grasp of others where you feel you have a better understanding.

Computer-based activities

You're likely to spend some of your time at university working on computer-based activities, but how much time will vary significantly from course to course. There are various pieces of specialist mathematics and statistics software that you might learn about during your degree, such as Mathematica, MATLAB, Octave and R, and you might learn about programming in a more general language such as Python. These tools are widely used in industry, so gaining experience in them during your degree can be useful when it comes to applying for jobs.

Coding skills are highly transferable: once you've learned to programme in one language, you'll have a head start on learning other programming languages.

Specialist mathematics and statistics software can transform your ability to tackle mathematical problems. For example, a computer algebra package can take care of routine but lengthy and fiddly computations, freeing you up to focus on the solving or modelling aspects of whatever problem you're working on.

You might also learn about a particular typesetting tool called LaTeX. This is widely used by mathematicians for writing maths because it's designed to handle mathematical symbols and equations. (This book was originally written in LaTeX, and the equations you'll see throughout were typeset using it!)

Assessment methods

There are many ways in which universities assess students' progress to determine degree classifications. You might sit written exams at the end of a module or at the end of the academic year. Your regular work during the term might contribute. You might have coursework projects to complete, either individually or as part of a group: these can be relatively short or come in the form of a longer dissertation that you'll work on over the course of a year. You might submit written work, or a logbook, or computer code and output, or you might have to give a presentation on your findings.

In some universities, your grades for the first and second years will be combined with your grade for the third year to give an overall degree classification (perhaps with a weighting so that the first year only counts for a small amount, with more emphasis on the third year). In others, each year is assessed separately (and recorded on your transcript). This is not necessarily something to worry about when choosing a course, but it's useful to be aware that it varies.

When researching the types of assessment on offer at university, you might want to find out how often you'll receive feedback on your work (how often will it be marked?). For instance, in some universities, your regular work during term time may contribute a small or large percentage to your overall mark for the module, while in others, you may be set regular work and receive feedback on it to help you improve, but these marks will not count towards your overall grade.

Department size and structure

Maths departments vary in size, in terms of numbers of staff (how many faculty are there?) and of numbers of undergraduates. You might like the idea of being in a smaller department, where you can get to know everyone well. Or you might prefer to be in a larger department, where there are potentially more options regarding what you can study.

A research-intensive department might have large numbers of research students (working towards their PhDs) and postdoctoral researchers, who might work with faculty to deliver undergraduate teaching. A department with more of a teaching focus might have faculty who spend more time teaching and less time on research. In any department, you can expect the people who are teaching you to have a thorough knowledge of their subject and to care about supporting your learning.

Maths societies

Depending on your interests, you might also want to find out whether the universities you're considering have maths societies. These might organize activities ranging from regular talks from professional mathematicians to hosting careers events. Some maths societies focus more on bringing maths students together for social activities, which is a great way to meet people and learn at the same time.

Academic maths support

During your degree, you'll require individual academic support at some point. Students making the transition to university have come from different backgrounds and studied different topics, so some individual attention can help to fill in any gaps left by previous experience. In addition, students find different topics challenging at different points throughout their degree, and it's useful to know that support is available when you're feeling overwhelmed.

Maths departments make support available in different ways. They might offer teaching in groups that are small enough to give you the opportunity to ask individual questions. They might have staff office hours or organize drop-in sessions, where you can go along to ask your lecturer or tutor for advice.

Students often find that working with their peers can be helpful too, both for making academic progress and for feeling connected to those around them.

Some universities have mathematics support centres, or even 'maths cafés', where students can work together and ask for advice from more senior students or from staff. Other universities have dedicated spaces in their departments where undergraduates can work between teaching sessions, either individually or collaboratively.

Perhaps the most important message of this chapter is that maths degrees vary widely, so there's a good chance you can find a course that suits you. As we've seen, maths degrees have differing content, and the modes of teaching and assessment differ too. You should be able to find information about all of these aspects from university websites, to help you make an informed choice before applying.

In Chapter 3, we'll address some questions you might still be asking yourself about whether a maths degree is right for you.

CHAPTER 3

Who can study maths?

THERE'S NO SUCH THING AS a typical maths undergraduate. There's a lot of variety in the maths degrees on offer (and in the options available within those degrees), and there's even more variety in the people studying to achieve them. Some students come to a maths degree straight after school, having known for years that they wanted to study maths at this level. Some students make the decision to study maths shortly before submitting their UCAS form, having agonized over a choice between subjects. And some students come to a maths degree a little (or a lot) later in life, having perhaps decided to change career or having found a new interest in maths.

In this chapter, we'll tackle a few questions that you might have about the kinds of people who study for maths degrees.

Do I need to have a career plan?

I've had students who started their maths degree with a clear plan: they wanted to go into school teaching or into investment banking, and, three or four years later, have done exactly that. I've had students who started with no idea what they wanted to do, just enthusiasm about maths, and this has evolved into a clear focus that has led them to software programming, data analysis for an environmental consultancy and transport planning. And I've had students who were dead set on a particular career path but have been inspired during their studies when they encountered an area of maths they didn't even know existed, whose career plans have taken a sharp swerve as a result: they're no longer going from their degree straight into finance, but continuing on to a PhD in mathematical biology, for example.

One survey of over 200 maths undergraduates[6] asked why they had chosen to study maths. Of those who took part, 74% said that the fact that they enjoyed maths or were good at it was a factor in their decision, whereas 32% mentioned careers (students could give more than one reason, if you're wondering why those numbers add up to more than 100%!). Among those who cited careers as a factor, the responses were more about having good career prospects than about wanting a specific career (although a few students did choose maths with a specific career in mind). If you're considering maths for career reasons, then, you're definitely not alone, but don't worry if you don't have a particular career plan in mind. If you're considering it because you enjoy it and do well at it, well, you're in the majority!

Some maths graduates will go into jobs that have existed for a long time (but are nevertheless evolving as the world around us changes): teaching, accountancy and banking, actuarial work and risk management, industrial modelling, software development, university research, and so on. We'll meet these in Chapter 4. Others will go into roles of a sort that have existed for only a few years: in data science, in machine learning or in artificial intelligence, for example. Others still will be involved in creating new roles as the impact of maths on society continues to expand. All this means that the flexibility of a maths degree is helpful: by the time you graduate, the jobs market will have shifted a little, and you'll have more idea of what kind of work is attractive to you and where your strengths lie, but your qualification and the skills you acquire while studying will be valuable, whichever direction you choose.

What should I factor in?

The following are a few questions you might want to ask yourself that could help you to form a better idea of where to start your studies – even if you end up somewhere completely different!

▸ Do I have a sense of the sort of job I'd like to take on (in which case, you might want to study the mathematical ideas or machinery relevant to that role)?

▶ Do I have no idea what I want to do workwise (in which case, you might want to study particular mathematical ideas that catch your attention, or that you want to explore further)?

▶ Do I enjoy abstract mathematics, or do I prefer more concrete ideas?

▶ Do I like the idea of having a broad overview of a range of ideas from maths, or would I prefer to specialize in a particular area and study this in depth?

▶ Do I enjoy working collaboratively with other students, or do I prefer to work alone?

▶ Do I like working with pen and paper (or pen and whiteboard), or do I prefer more computer-based mathematics?

▶ Do I excel in written exams, or do I perform better in sustained project work over a period of weeks or months?

As previously discussed, the variety that a maths degree offers means that no matter your preference, you're likely to be catered for. It's important to remember that whoever you are, whatever your background, whatever your interests in the mathematical sciences, whatever your career aspirations for the future, maths needs you.

Do I have to love maths?

Some maths students really, really like maths, and that's great. At university, they're in an environment where they can immerse themselves in studying the subject they love, join the maths society (if there is one) and share their excitement about maths with other students. If that sounds like you, then you're going to love doing a maths degree! However, not everyone is *quite* so passionate about maths, and it's important to note that you don't have to be a maths devotee in order to be a successful maths undergraduate.

If you're going to take maths at university, then you're going to spend a lot of the next three or so years studying the subject, so it'll definitely help if you like it. Having said that, you're not expected to spend every waking minute thinking about maths. You'll have plenty of time for a social life, sports, clubs and societies, and your hobbies. You might be expected to do some academic work over vacations, but not as much as during term time. You might have a part-time job or caring responsibilities. All of these will complement your maths studies.

You might find that you enjoy some of your modules but aren't so keen on others. That's very natural: we all have different tastes and preferences. If the degree programme gives you scope to choose some options, particularly in later years, then you'll be able to tailor your studies to suit your interests.

What if I can't decide?

Students sometimes face a difficult decision between two (or even more) subjects: they like both, and they're doing well at both, so how should they decide what to do at university?

One possibility might be to do a joint degree, as mentioned in the previous chapter, which will allow you to pursue your interest in maths alongside another subject. If a joint degree isn't an option or doesn't feel right for you, then here are a couple of other ways that people sometimes think about how to choose between maths and another subject.

Some students think about whether it'll be possible for them to maintain their interest in the subject in another way, that is, not as part of their main degree. A student who's interested in maths and languages might choose a joint degree, or they might choose to study a maths degree that allows them to take one or more modules in languages, or they might choose to study a maths degree and keep up their language study informally on the side. Sometimes students decide they will focus on maths in their degree and pursue their interest in history or English or another subject as a hobby, through reading and watching documentaries. It might, for example, be harder to study history as a degree while keeping maths as a hobby.

Another way to think about your choice is to picture, for each option, what sort of activities will be involved on a day-to-day basis. For example, if you're very interested in theoretical physics, then you might be considering a joint maths and physics degree – but you might also be considering separate degrees in maths and in physics, each of which could naturally lead to specializing in theoretical physics in later years (depending on the precise degree programme). In a physics programme, you would probably be expected to do some practical experimental work in a laboratory, whereas this is unlikely to be part of a maths degree. Some students find that this way of thinking can help them to focus their choice: maybe they love laboratory work and wouldn't want to lose that, or maybe they'd be only too pleased never to have to do a practical again! When you're researching degree courses, you can try to find out about the sorts of activities that will be involved and what balance there is between them. How much time will you spend in lectures, in tutorials or classes, working on problems sheets, working on computer-based activities, working by yourself, working in groups or working on projects? Are there activities that you'd be glad to see the back of, or ones that you'd really miss if they didn't feature in your degree?

For some students, this can really help them to focus on the right programme for them.

Do I have to be a genius?

Some subjects have a reputation. You might think that you need to be a genius in order to do them, or that you need to have been born with a talent for them. It seems that maths is often placed in this category, but it shouldn't be. In maths, as in so many areas of study and of life, by working hard, you can improve and progress.

Whether you have a 'growth mindset' or a 'fixed mindset' has been shown to have an effect on academic achievement in work by Carol Dweck (a professor of psychology at Stanford University),[7] Jo Boaler (a professor of maths education at Stanford University)[8] and others.

Of two minds

People with a fixed mindset about maths think that each person has a fixed level of maths ability. There are things they can do in the subject, but eventually they'll reach the limit of what they can achieve. People with a growth mindset about maths, however, believe that people can continue to grow and develop their mathematical capabilities.

These different perspectives can affect how a student views making a mistake on a maths problem. For a student with a fixed mindset, making a mistake is a cause to wonder whether they're clever enough to continue studying maths: perhaps they've reached the limit of what they can possibly achieve and should therefore quit. For a student with a growth mindset, making a mistake is seen as an opportunity to learn, to understand why their work was not correct and to make new connections that will help them in the future.

What might a growth mindset mean in maths?

Let's illustrate this with an example centred on a topic called analysis, which you might meet during your maths degree.

Many students don't meet analysis at school or college, so they can find it a challenging new topic at university. In a nutshell, analysis is about the rigorous foundations of calculus. Some questions that an analysis course might seek to address are the following:

- ▶ How do we define what it means for a sequence or series to converge?

- ▶ Given a function f, what does it mean to say that $f(x) \to 3$ as $x \to 0$?

- ▶ How do we rigorously define differentiation and integration?

The way in which we learn about calculus mirrors, in some respects, the history of the subject. At first, we learn what differentiation means, and how to do it, and similarly for integration. Once we've added the necessary mathematical tools to our toolkit through study and practice, we can use them to answer increasingly sophisticated mathematical problems, and to model real-world problems. This is what mathematicians were doing from the seventeenth century onwards. By the nineteenth century, some mathematicians were becoming increasingly concerned with setting out the rigorous foundations of the subject. Instead of relying on intuitive or imprecise notions of convergence and limits, they worked to give formal definitions from which they could prove careful theorems. Their work forms the basis of the modern study of analysis.

For example, you might already have some intuition as to what it means for a sequence to converge (as in the first question above).

Here are some examples of sequences:

(i) $1, \frac{1}{2}, \frac{1}{3}, \frac{1}{4}, \frac{1}{5}, \ldots$

(ii) $1, -\frac{1}{2}, \frac{1}{4}, -\frac{1}{8}, \frac{1}{16}, \ldots$

(iii) $\frac{5}{2}, \frac{7}{3}, \frac{9}{4}, \frac{11}{5}, \frac{13}{6}, \ldots$

(iv) $1, 2, 3, 4, 5, \ldots$

(v) $1, -1, 1, -1, 1, \ldots$

Which of these sequences would you say converge, that is, tend to a limit?

Based on mathematical instinct, it looks as though sequence (i) tends to 0. Meanwhile, sequence (ii) is more subtle, because its terms are alternately positive and negative; however, it seems that it also tends to 0. It's harder to see with sequence (iii). Illustrating it graphically might help, or rewriting it as

$$2 + \tfrac{1}{2}, 2 + \tfrac{1}{3}, 2 + \tfrac{1}{4}, 2 + \tfrac{1}{5}, 2 + \tfrac{1}{6}, \ldots$$

might suggest that it tends to 2. Depending on your perspective, you might feel that sequence (iv) doesn't converge, or you might say that it tends to infinity. And while sequence (v) oscillates in an interesting way, it doesn't seem as though it converges.

As we can see from the examples above, there are many types of sequence and they behave in many different ways. How, then, can we write down a formal definition of convergence: a definition that can cope with all of these different behaviours? The standard definition of convergence is something like this:

We say that a sequence (x_n) of real numbers tends to a limit L if for all $\varepsilon > 0$ there is some N such that if $n \geqslant N$ then $|x_n - L| < \varepsilon$.

That may look pretty scary at first glance, and unfortunately this isn't the place to unpick what it means (but if you're keen, you can try it out on the sequences above). We're just using it as an example of a new idea to explore.

Did you look at the above and think 'I'm not clever enough to understand that', or did you think 'I have no idea what that means, but it's intriguing and I'd like to learn about it'? If you were in the former camp, don't worry; we've all experienced being daunted by a mathematical challenge, and feeling this way from time to time doesn't mean that you shouldn't do a maths degree. However, you'll get a lot more out of your studies if you embrace the latter mindset and feel positive about embracing this sort of challenge with appropriate support. This is more of the approach of someone with a growth mindset, and it will serve you well in university and beyond.

Like other aspects of maths, problem solving is a skill that you can develop, and studying for a maths degree will help you to do exactly that. In order to strengthen your problem-solving muscles, however, you'll need to practise on problems that you don't already know how to solve. By working on these, whether by yourself or with others, and getting feedback from your teachers, you'll develop invaluable skills.

In addition, you'll experience teaching methods in a maths degree that you won't have met at school or college, and you'll learn about new ideas, techniques and approaches. Often, meeting these new areas of maths will mean learning new terminology and new notation, and you'll have to get your head around these before you can fully understand a topic. By having a growth mindset – by being open to new challenges and receptive to new ideas – you'll be equipped to face these challenges and free to develop into a successful maths graduate.

It's all about the attitude

The bottom line is this: you don't have to be a genius to study maths, you've just got to have the right outlook. Your degree course will be designed to support you and your fellow students on this journey, and if you work hard, embrace challenges, learn from your mistakes, listen to your teachers and colleagues, play around with ideas, and take feedback on board, then you'll do well.

Your maths degree will help you to build up your study skills, such as managing time effectively and acting on feedback, but it will also assist you in developing attributes such as resilience, curiosity and perseverance. These skills and attributes will be important to your future success.

Do I have to fit a stereotype?

When you picture someone who studies maths, what do you see? There are plenty of stereotypes about mathematicians and maths students, and, like stereotypes in other areas, they're neither accurate nor helpful. Some of these are related to various forms of diversity.

Gender

Historically, maths has had a significant gender bias. If you think about famous mathematicians of the past, you probably think mostly of men. This has changed in recent years, and it's continuing to change: there are growing numbers of women and non-binary mathematicians.

The Higher Education Statistics Agency (HESA) recorded 37,620 undergraduates in the mathematical sciences (including full- and part-time students) in 2017/18.[9] Of these, just over 37% were female.

The gender balance among undergraduates varies between university maths departments, but universities are actively engaged in seeking to improve it, both among students and among staff.

You'll find that some university departments have achieved an Athena SWAN award,[10] at bronze, silver or gold level. To obtain such an award, a department must analyse its policies and processes and the data about its current situation in terms of gender equality, and it must demonstrate that the department's working to improve it. The slogan of the Athena SWAN Charter is: 'recognising advancement of gender equality: representation, progression and success for all'.

There's a range of actions that university maths departments are taking to work towards gender equality. You'll find social and networking events for women and non-binary staff and students, mentoring, student-led societies, funding to support those with caring responsibilities, and national events for women in mathematics.

At a national, and even international, level, the London Mathematical Society's (LMS's) Women in Maths Committee[11] organizes events and supports maths departments to progress towards gender equality, while European Women in Maths[12] brings together mathematicians from many countries, to promote and support women in mathematics.

Ethnicity

There are other underrepresented groups in mathematics too, for example, based on ethnicity. Again, this issue has arisen from a complicated mix of historical and societal factors, and universities are keen to address the imbalances. The high-level HESA data classifies students as White, Black, Asian, mixed, other and not known. For the 2017/18 cohort of mathematical sciences undergraduates, 75% identified as White, 3% Black, 15% Asian, 4% Mixed, 1% Other and 1% Not Known. Across all subjects for 2017/18 undergraduates, the figures were 75% White, 7% Black, 11% Asian, 4% Mixed, 2% Other and 1% Not Known. If you're interested, a more detailed breakdown by ethnicity is available from HESA.[13]

There's now a Race Equality Charter,[14] the goal of which is 'improving the representation, progression and success of minority ethnic staff and students within higher education', and some universities hold, or are working towards, awards indicating their commitment to this aim.

You'll find activities within universities to 'decolonize the curriculum', an institutional emphasis on more diverse role models and imagery on the walls as well as student-led societies and projects. One interesting project is Black Mathematician Month, launched by a group of mathematicians to showcase black mathematicians of the past and present, and to explore ways to increase diversity in the future.

Disabilities

A significant number of maths undergraduates have a disability of some sort and thrive in their studies. The HESA data for undergraduates[15] shows that of the 37,620 mathematical sciences students in 2017/18, just under 12% were known to have a disability (compared with 14% across all subjects).

Universities have extensive experience in supporting these students and will often have a disability advisory service, the purpose of which is to offer advice, guidance and support to students and to the staff working with them. There may be specific funding that you can access, if you require additional provision of any sort, and the disability advisory service would be able to help you to apply for this if appropriate.

Universities are increasingly exploring the 'inclusive' model of education, where teaching and assessment are designed from the outset to include as many people as possible, rather than needing to make specific adjustments for individuals (although sometimes there will be a need for bespoke adaptation, and universities are obliged to make 'reasonable adjustments' to allow all students to access courses).

Over the years, I've taught students with physical disabilities, students with long-term mental health conditions and neurodiverse students. Some students knew before they came to university that they might require adjustments or additional support in some form, and they were able to work with the disability advisory service to make arrangements before their course started. Other students' circumstances changed during their degree, or they received a new diagnosis of a long-standing condition, and we were able to put appropriate support in place in response.

If this is an aspect of university study that concerns you, then you can research and contact disability advisory services at individual universities to learn more about the provisions that are in place. You might also want to look up the STEMM Disability Advisory Committee, which brings together representatives from key bodies in science, technology, engineering, mathematics and medicine to support 'disabled workers, current and aspiring disabled students and their teachers'.

Although maths wasn't diverse in the past, that's changing, and it needs to continue to change. It's important that maths graduates come from diverse backgrounds and bring diverse perspectives to their work. The IMA and the LMS have put together profiles of a range of mathematicians from diverse backgrounds, and if you'd like to be inspired by their stories, they're available at https://ima. org.uk/about-us/diversity-statements/ (click on various aspects of diversity to reach the interviews) and https://www.lms.ac.uk/ careers/success-stories.

Maths needs you

Even if you don't know exactly where you're heading, you're not maths-obsessed, you don't think you're a genius or you don't fit the 'stereotype' of a maths student, there's a place for you in maths. All kinds of people study maths, and maths needs all kinds of people.

A maths degree can also lead to all kinds of graduate destinations. Do graduates from a particular programme tend to continue on to further study, or to move into the world of work? What jobs do maths graduates do, and how employable are they? These are the questions we'll explore in the next chapter.

CHAPTER 4

What do maths graduates do next?

AS PROFESSOR PHILIP BOND, AUTHOR of *The Era of Mathematics: An Independent Review of Knowledge Exchange in the Mathematical Sciences* (2018), explains:[16]

> *The mathematical sciences deliver significant social and economic impact in the UK. Mathematical tools and techniques lie at the heart of numerous industries, ranging from financial services to the special effects and computer-generated imagery (CGI) used in the film industry, underpin much of the technology used in national security and defence, and are now essential in the life sciences. Medical advances increasingly rely on mathematical data analytics, machine learning and process modelling, while medical imagers such as magnetic resonance imaging (MRI) scanners use algorithms directly derived from mathematical methods. Engineering and material sciences remain heavy users of mathematical methods, allowing the UK to remain a leader in numerous advanced engineering fields such as aerospace and Formula 1 motorsport.*

Maths graduates are generally highly employable and, as Bond makes clear in the above statement, have many possible routes open to them. Do you already have a plan for how you'd like your career to develop?

Some people choose to study a maths degree because they know it'll help them progress towards a particular career, and there are many jobs for which a maths degree is the perfect preparation. Other people opt for a maths degree simply because they enjoy maths, and have been doing well at it, without knowing what they want to do when they graduate. A maths degree is ideal for people in the latter situation because it's so immensely versatile.

If you already know what your career plans are, then you can check that a maths degree will get you to where you want to be. But don't worry if you don't know what you want to do: plenty of maths undergraduates are in this situation. One survey of over 200 new maths undergraduates (studying at several UK universities) found that only 43% of them had any career plans at the start of their degrees.[17] If you're interested in maths, then a maths degree can be a rewarding and satisfying way to spend three (or more) years, helping you towards a future career and increasing your employability while keeping your options open.

There are plenty of jobs where the subject knowledge gained during a maths degree is important, but there are even more jobs where the broad range of skills and expertise obtained by maths graduates is what's valued by employers, rather than an in-depth knowledge of any particular topic. This means that embarking on a maths degree because you find maths interesting and satisfying to study is a perfectly sensible choice, and you can do so safe in the knowledge that there will be many doors open to you when you finish your degree.

This chapter will give you some insight into the variety of careers in which maths graduates find themselves: you might be surprised to learn that the title of 'mathematician' often doesn't feature! It'll also provide information on salary expectations and offer some ideas on how to enhance your employability while still at university.

The value of a maths degree

Society needs doctors, nurses and other medical professionals to care for us when we're ill. It needs architects and engineers to design buildings and structures that enhance our lives. It needs lawyers and police officers. It needs social workers, hairdressers, chefs and plumbers. It also needs mathematicians: people with advanced mathematical training. We're facing some big challenges in modern society – challenges relating to climate change, health care and the economy, to name just a few – and mathematics has an important role to play in addressing them. You'll find more about that in Part II of this book.

Maths graduates are, of course, highly numerate, but they're also good at solving problems, at logical reasoning, at expressing complex ideas in precise ways, and at using a computer to tackle challenges. These are skills that employers really value.

In October 2019, the QAA subject benchmark statement for mathematics, statistics and operational research said:[18]

Graduates of mathematics, statistics and operational research courses have an extremely wide choice of careers available to them. Employers greatly value the intellectual ability, rigour, logical thinking and abstract reasoning that graduates acquire, their familiarity with numerical and symbolic thinking, and the analytic approach to problem-solving that is their hallmark. These skills, when developed alongside more generic skills (such as communication and team-working skills), make mathematics, statistics and operational research graduates highly employable.

If you're studying maths at the moment, then well done! You're already on the path to developing these skills – even if you weren't aware of it. And studying maths at university will help you to develop them even further.

What do maths graduates do?

Those who study medicine might well go on to become doctors. Those who study law, lawyers. Those who study engineering, engineers. There aren't so many people whose job title is 'mathematician', so it can be hard to know what people who study maths end up doing. Here's the good news: the reason for this is the huge breadth of opportunities available to maths graduates; it's nothing to do with there being a shortage of career openings. In addition, a maths degree can give you the flexibility needed to shift direction in your career. Many people change roles multiple times, or even choose to head down a completely different path, during their working lives. The skills you acquire during a maths degree can give you the versatility that makes this possible.

In fact, like the medicine, law and engineering graduates mentioned above, some maths graduates choose to train as doctors, lawyers or engineers after completing their degrees, drawing on the skills and experience they've obtained by studying maths.

While you might bump into a maths graduate in almost any industry, there are a few career paths to which students in this field are particularly drawn. The following is by no means a complete list of job options, but it should hopefully give you some idea of the areas where maths students flourish.

Finance

Perhaps unsurprisingly, many maths graduates choose to work in finance roles. These include accountancy, investment banking and actuarial work, to name just a few areas.

We briefly met the area of actuarial science in Chapter 1, and the role of an actuary came up again in Chapter 2, because some degrees include modules that are accredited, meaning that completing the module successfully gets you exemption from one or more of the exams needed to become a qualified actuary. Actuaries analyse and measure risk relating to pensions and to insurance, for example. There's a Government Actuary's Department, which helps policymakers understand and take account of risk when making key decisions. The 'further reading' section includes a range of sources where you can find more information about this and other careers that might interest you.

Some roles in finance are more customer-facing, with an emphasis on understanding and presenting complex data as well as getting to grips with clients' needs. If you're a maths graduate with excellent communication skills, then a role like this could allow you to combine your technical expertise with helping people. Others are backroom positions that may involve designing strategies, modelling scenarios, interrogating data, finding technological solutions or all of the above. Maths graduates are known for the precision, logic and rigour with which they approach tasks as well as their ability to deploy problem-solving skills. They're also well placed to quantify risk in order to manage it appropriately. Having some experience of coding can be an advantage in this sector too.

Data science and machine learning

Today, data science and machine learning are hot topics. The former is about analysing data to draw useful conclusions and is relevant in many areas of business as well as in medicine, health care, climate change and beyond. The main goal of the latter is to get a computer to learn what patterns and structures to look for in a set of data, rather than having a human tell the computer what to look for. This is proving increasingly powerful in analysing many sorts of data sets.

Those with maths degrees are in high demand in these areas, for roles such as 'data scientist' or 'data analyst', because the work draws heavily on sophisticated mathematical and statistical ideas. We'll discuss this further in Chapter 6, which looks at data science and machine learning in a bit more depth. Modern life allows organizations and companies to gather huge amounts of data, which, if analysed carefully, can transform their businesses or activities, in fields as diverse as genetics research and retail.

Software development

Maths graduates often go into software development roles, using their expertise related to computer programming. This might draw on particular coding experience as well as the problem-solving, logical reasoning and analytical skills of maths graduates.

Some maths graduates will have learned some coding as part of their degree, or even as a hobby. For some roles, you might need familiarity with a particular programming language, but often it doesn't matter which language(s) you've already learned: what's more important is knowing the general principles of coding, which will allow you to pick up another language more quickly. Some maths graduates even go into this sector without programming experience: their skills and expertise from their degree mean that they learn quickly when provided with training by their new employer.

Cryptography and cybersecurity

Continuing in the technical vein, private companies and government departments alike employ maths graduates to work in the areas of cryptography and cybersecurity. In a nutshell, cryptography is about secret messages: encrypting data to keep it secure, and decrypting it to see what has been transmitted. There's more about this in Chapter 6.

Interestingly, there are many connections between 'pure' mathematics and these disciplines, where number theory, abstract algebra and algebraic geometry can be applied to powerful effect.

Education

There's always a demand for more maths graduates to choose the teaching route, given how vital it is for our future that all children and young people have access to a high-quality (mathematics) education. This might not be the most lucrative career choice, but it can be hugely rewarding in other ways. If you're considering going into teaching, then you might want to undertake a joint degree that incorporates education alongside maths. Another option is to choose a maths degree that will give you the chance to take one or more modules in the area of maths education and, perhaps, to get experience in a school classroom context as part of this. There are various routes for maths graduates to get into teaching, whether via further study (a PGCE) or training on the job.

Don't forget, there are other careers within the education sector besides teaching: for instance, you could work for an organization that creates educational materials.

Research

While a lot of maths graduates go into employment once they've graduated, in many different roles and in many different sectors, some choose to continue their studies by doing a research degree (a research masters, or a doctorate – a PhD). They might then move on to a research position in industry, at a university or for the government. If you think you might be interested in mathematical research in the future, please be reassured that there are still many unsolved maths problems – and many more that haven't been asked yet!

Maths needs new people to find new answers and pose new questions. The examples we'll see in Part II show that maths is continuing to grow and develop, and this is only possible thanks to the creativity, innovation and determination of mathematicians and their collaborators. The people teaching you during your maths degree might be engaged in mathematical research of their own, continuing to push the frontiers of the subject and finding benefits in the mutual interplay of teaching and research.

It would be hard to overstate the variety of industries that benefit from the skills of maths graduates, and while the above categories are perhaps the best-known employers of maths students, they're definitely not the only ones.

Where do maths graduates work?

Many maths graduates work in statistics, mathematical modelling and OR in a range of industries as well as for governments, charities and non-governmental organizations (NGOs). While some will be employed by a company, others will decide to set up their own businesses. You'll find maths graduates working as civil servants, for banks and pension funds, and for supermarkets, Formula 1 teams, the NHS, pharmaceutical companies and environmental consultancies, all of which employ data scientists and operational researchers.

Those with mathematical modelling expertise can work in weather forecasting and climate science, for defence companies and in the transport sector. In 2020, some of this work grew in prominence as the role of maths in modelling and analysing the Covid-19 pandemic was brought to the attention of the public. The use of mathematical modelling to study the spread of disease is discussed in Chapter 5, along with other applications of maths in diverse situations. You'll also find more about the practical uses of maths in Chapter 6, including an application of OR to improve the system for kidney donations in the UK and examples of mathematical modelling being used in the fields mentioned above.

Finally, some maths graduates go into roles for which their maths degree is not a prerequisite, but where the knowledge they've gained during their degree is nonetheless useful and valued by employers. These roles are incredibly diverse, so it's difficult in this short book to give you a sense of the range of directions taken by maths students. See the 'further reading' section at the end of this book for a list of sources illustrating the various careers available to maths graduates.

What proportion of maths graduates get jobs?

The Higher Education Statistics Agency, the UK's official agency for gathering data related to higher education, conducts an annual survey of graduates six months after they've completed their degrees. Their 2018 report[19,20] shows that 62% of UK-domiciled maths graduates (having just completed a first degree) were working or combining work with study, and an additional 26% were in further study. (The remaining graduates were unemployed, retired or doing other activities, such as travelling or taking a career break to care for a family member.) The corresponding figures for all graduates of a first degree were 72% and 17%.

An additional HESA study[21] looked at what graduates were doing three and a half years after they completed their studies. In the case of people who graduated with a first degree in maths, at this stage (a few years after graduation) 88% were working or combining work with study (compared with 87% for all subjects), and an additional 9% were in further study (compared with 7% for all subjects), leaving just under 3% unemployed or doing other activities (compared with 6% for all subjects).

Salary

Different people have different priorities for their working lives. While money is just one of the factors you'll need to consider when deciding on your course of study, it's a factor that many people think is important. Thankfully, the evidence shows that maths graduates are among the more highly paid of those who undertake higher education.

According to recent data from HESA,[22] UK-domiciled graduates in the mathematical sciences who enter full-time paid work in the UK are among the highest paid (based on median salary), just behind graduates in medicine and dentistry, veterinary science, and engineering and technology. Their salaries are also above average for all subjects, including all science subjects. For example, HESA data for the 2016/17 academic year show that maths graduates going into full-time work received a median salary of £25,000, with 25% earning £30,000 or more. By contrast, the median salary for graduates in all subjects for the same academic year was £22,000, with 25% earning £26,000 or more.

A HESA longitudinal survey[23] looked at those who graduated with a first degree in the 2012/13 academic year, to learn what they were doing three and a half years later (in 2016/17). The median salary for maths graduates (in employment, but not including those in self-employment) at this stage was £30,000 (compared with £26,000 for all subjects), with 25% earning £38,000 or more (compared with £32,000 for all subjects). Notably, 90% of maths graduates were earning £21,000 or more; to put that in perspective, only 74% of graduates overall were earning at this level.

In terms of the median salary, maths graduates were ranked joint-fourth out of all subject groups, again behind only medicine and dentistry, veterinary science, and engineering and technology graduates. The other subject group with graduates in fourth place was architecture, building and planning.

Looking at people's careers a few years down the track, a study by the UK government's Department for Education in partnership with the Institute for Fiscal Studies used the Longitudinal Educational Outcomes data set to explore the impact of higher education on people's earnings at age 29. This study listed maths among the top three subjects for mean earnings (for both men and women), along with medicine and economics.[24]

That's not to say that every maths graduate earns this much: some career paths are less well paid but will offer other benefits. Nevertheless, if a high salary is a priority for you, then studying maths can be a good option.

It's worth thinking carefully about what you hope to get from your working life before you embark on a career. Some people want to be earning a large amount, while others will have lower financial expectations and will prioritize having a great work–life balance or working in a particular geographic location. Some people want jobs where they'll be making a difference to their local community or to wider society, while others want a challenging role or particular career progression possibilities. The fact that a maths degree is useful preparation for a variety of roles means that no matter what your individual priorities happen to be, studying maths can help you to focus on and achieve them.

Enhancing your employability

While obtaining a maths degree will open doors to many forms of employment, there's plenty you can do while you're still studying to make these doors a little easier to walk through. Some university maths departments are particularly proactive about offering activities relating to careers and employability, either as part of a formal degree programme or as an optional extra for those students who are interested. Don't forget that your university might also have a central careers service that can provide support and advice relevant for maths students.

Academic work

You shouldn't just focus on activities and material that are explicitly labelled as being about personal development or employability (important though these are), because regular academic endeavours can help you to develop valuable skills as well. For instance, as we mentioned in an earlier chapter, taking the opportunity to work as part of a group can enhance your skills in collaboration and teamwork. In addition, undertaking a thesis, dissertation or other sustained project will help boost your research skills and give you a chance to add planning and carrying out independent work over a period of time to your skill set. And when you write up your project findings in a report, or prepare a presentation on them, you'll be honing your communication skills: a favourite among employers.

As we saw in Chapter 1, maths degrees can be thought of as lying on a spectrum from 'theory-based' to 'practice-based'. This means that some will seem more immediately relevant to the world of work than others, but whatever style of course you choose – even in your day-to-day work on problems sheets or projects, or by talking to your tutors and fellow students in teaching sessions – you'll be honing your ability to articulate complex ideas in a precise, rigorous and logical fashion. As you tackle your academic work, you'll learn more about how to learn, how to grapple with material that you find challenging at first, how to use your creativity and repertoire of techniques to solve mathematical problems, and how to analyse a scenario in order to model it mathematically. By the end of your degree, you'll be better equipped to teach yourself a topic independently, which is a valuable life skill.

Learning to use a computer to tackle a mathematical problem will give you experience with coding and using specialist software, which is useful in a wide range of careers. Your job might not use the same software or programming language that you have worked with during your degree, but the skills and general principles you'll absorb will be very transferable: once you've learned one language, it's much easier to pick up another.

You might already have a clear sense of what you'd like to do after your degree, in which case you can be alert to relevant opportunities during your course, to help you populate your CV with appropriate skills and experiences. Even if you don't know what you want to do when you graduate, it's still worth being proactive: explore the activities on offer and think about developing your portfolio, as many skills and experiences are valuable in a wide range of areas.

Internships

Employers will often target universities to recruit students for internships as well as for future employment. Indeed, doing a summer internship can sometimes lead to a job offer after your degree. These internships give you the chance to try out a job to see whether it's a good fit for you and allow a prospective employer to decide whether or not they want to hire you. You might want to explore whether any paid internship opportunities are being advertised within your department. You might also be interested in whether the department offers funded summer research projects, which can give undergraduates a taste of mathematical research.

Some students even spend a whole year working in industry as part of their degree. This allows them to get a real insight into a possible future career path while building skills and experience that will help with the remainder of their degree programme and with applying for future jobs. A placement year isn't an option in all degrees, so if this is important to you, it's a good idea to focus on the programmes that include it.

Extra-curricular activities

Your extra-curricular activities can also help you to develop attributes that will be useful in your subsequent career. Perhaps you'll be on the committee of a student society or you'll join a sports team or musical group: this can be key in developing leadership, teamwork and organizational skills. Perhaps you'll have a part-time or vacation job, or undertake voluntary work in the local community, where you can demonstrate and expand your portfolio of skills. It's important that you don't overlook the value of all this in increasing your employability. By being aware of the skills you're developing as a student, you can seek out further opportunities to enhance them even more. You'll also know to share them with prospective employers when applying for jobs or further study, and they'll help you think about your strengths if you're creating your own business.

You don't need to know where you're going

A maths degree is a very versatile qualification and can leave you well placed to pursue a wide variety of possible career paths. You might already have an idea of what you'd like to do when you've completed your studies, but it's also perfectly fine to start a maths degree just because you find maths fascinating and satisfying to study, and to wait until your degree is underway (when you'll have more information about your strengths and interests) to make longer-term plans.

You might think that your choice of future job should inform which topics you study at university. It's true that you might choose a particular subject because it relates to your intended career, especially as some modules are accredited to lead to professional qualifications after graduation. And, naturally, someone with interests in a particular field is likely to choose modules relating to that area. A student who wishes to become a school teacher, for example, might choose to take a maths education module in order to study theories on how people learn, to gain relevant experience and to learn more about their planned career. A student looking to work in the financial sector might choose modules relating to financial mathematics. However, for many jobs, your precise combination of modules won't be important. Often employers don't expect you to have expert knowledge of a particular mathematical topic and will instead be looking for the broader skills you've acquired by studying maths. In addition, some students won't decide on a particular career path until quite late into their degree or, indeed, will change their minds significantly during the course of their studies, having encountered a broader range of exciting options.

All this means that many students choose their modules based on what they find interesting and where they feel their strengths lie. Happily, there's usually enough flexibility in a maths degree to let students find topics and modules that appeal to their unique interests.

While there's no time like the present to start researching current jobs held by people who studied maths at university (the 'further reading' section includes suggestions of resources to help with this), remember that in five years' time there will also be new types of role and industry: a reflection of the increasing demand for tools and techniques from maths, statistics and OR in a modern world with new technology and an ever-increasing reliance on data.

The rest of this book will give you even more reasons to study maths as well as a more in-depth look at some of the topics you're likely to meet during your degree and in your subsequent career.

PART II

MATHS IN ACTION

NOW THAT WE'RE FAMILIAR WITH what a maths degree involves and where it could take you career-wise, the next three chapters will look in more depth at some of the topics you might study in a maths degree, showcasing intriguing theoretical ideas and applications that have a real impact on people's lives.

Chapter 5 delves into three subjects that deserve special attention: differential equations, linear algebra and complex numbers. These are emphasized because they occur in pretty much all maths degrees: the QAA's benchmark statement (see Chapter 1), which basically sets out the national expectations for maths degrees and maths graduates, highlights calculus and linear algebra as topics common to all maths degrees. This chapter will give you a taste of what these subjects are about and how they can be applied.

Chapter 6 will look at a series of case studies to provide you with a glimpse of the real-world settings in which some key mathematical techniques are important. Each one will look at a mathematical idea or tool you might come across in your undergraduate studies, as well as an example of how this approach can be used in practice. The goal is to give you a flavour of how these tools are employed, rather than to go into exhaustive detail.

Following this, Chapter 7 will offer a few case studies illustrating some themes in fundamental mathematics, also called pure mathematics. We'll see how mathematical reasoning can be used to tackle intriguing theoretical questions and can lead us to surprising conclusions.

It's important to remember that the fields of applied mathematics and pure mathematics are not mutually exclusive, though: many of the tools and techniques we'll meet in Chapters 5 and 6 were originally developed with no particular real-world application in mind (or, at least, not the application(s) they're now used for), while those in Chapter 7 can be put to good use beyond theory.

CHAPTER 5

Applications of key mathematical ideas

MATHEMATICS IS A POWERFUL AND pervasive language for describing, analysing and making sense of the world around us. During your maths degree, you're likely to study mathematical tools that are effective in a wide range of contexts, and to explore applications of mathematics to problems from a variety of disciplines.

Part of the versatility of maths is that once you've got a well-stocked toolkit, you'll be able to tackle the issues that arise from different applications. While your tools will need to be tailored to the individual problem, many methods are very general and can be deployed in lots of different situations. Sometimes the tools involved will be relatively straightforward to use, but setting up the problem in a way that allows us to use mathematical ideas might pose a significant challenge. Other times it could be that we need tools that are more sophisticated or that have been customized or developed in new directions to solve the issue at hand.

As we've seen in Part I, maths degrees vary, so the tools you'll learn about and the approaches taken will differ between degree courses. However, topics such as differential equations, linear algebra and complex numbers are so fundamental that they're typically explored in some way in all maths degrees. These university-level topics build on and extend areas of maths you might have studied at school or college, and as you'll discover they're extraordinarily powerful and vital for modern society.

In this chapter, we'll see some applications of differential equations, linear algebra and complex numbers, which should give you a glimpse of where your mathematical studies could lead.

DIFFERENTIAL EQUATIONS

Modelling the spread of disease

Maths saves lives. This was brought into sharp focus in 2020, when mathematicians and statisticians played a key role in the worldwide effort to understand and contain the spread of Covid-19, the disease caused by the most recently discovered coronavirus. You might have noticed mathematical modelling being discussed in TV news reports as it became clear that this real-world application of maths was crucial in informing government policy.

This is not the first time that maths has played a part in tackling the spread of disease. For example, in 2014–16 there was a major outbreak of Ebola in West Africa, while in 2001 foot-and-mouth disease infected farm animals across the UK; mathematicians and statisticians were important in handling both crises. At first sight, it might not be obvious what a mathematician or a statistician could contribute to epidemiology (the study of how and why diseases spread). However, monitoring and controlling a virus raises a lot of questions that mathematicians and statisticians are equipped to answer. If there's a vaccine, how should it be deployed? How many people need to be vaccinated, and who should be prioritized? If there's no vaccine, what preventative measures need to be taken (such as restricting people's movements and closing businesses)? How long will the disease remain in the population?

Finding the solutions to such problems through mathematical modelling and forensic analysis of data generates vital information that policymakers, politicians and medical professionals can use.

Now that we know why mathematicians and statisticians are vital to epidemiology, it's time to find out how they put their skills to use. How can we model the spread of disease using maths?

The SIR model

One way in which mathematicians seek to understand the spread of disease is by using the SIR model, which consists of a set of differential equations. Mathematicians can solve these equations to predict how the disease will unfold over time. They often use data from past epidemics, such as the outbreak of influenza between 1918 and 1920 that killed tens of millions of people around the world, to test the accuracy of their models. If the model's prediction closely matches the historical data on what actually happened, then we'll know this model is more likely to provide an accurate forecast.

Since different diseases behave in different ways – for example, some are more infectious than others – the model needs to take this into account. There's a generic collection of equations that need to be tailored to each particular outbreak of disease: we do this by specifying some numbers, known as parameters, which capture factors such as how infectious the disease is.

Here's the approach of the SIR model. We think of a population (of N people) and divide this into three categories. At any given time, there are S 'susceptible' people, who haven't been infected yet but could be; I 'infected' people, who can spread the disease further; and R 'recovered' people, who have already had the disease, are no longer infectious and are now immune. (In the case of a fatal disease, fatalities would be included in this third category, and for this reason R sometimes stands for 'removed' rather than 'recovered'.) While each of the three numbers S, I and R changes over time, in a way that's described by the equations that make up the model, there's one constraint: they must always add up to give the total number of people. That is, $N = S + I + R$.

Now we need to use some maths to capture how the quantities S, I and R are interconnected. At the start, before the disease arrives, perhaps everyone is susceptible, or perhaps most people are susceptible, while some have natural immunity. Over time, some susceptible people will move into the infected category. This model is based on the idea that the number of people who become infected will depend on the number of people who are already infected, and are therefore infectious, and on the number of susceptible people.

If there are many infectious people and many susceptible people, then we can expect many more people to become infected. How infectious the disease is, which is a constant parameter, will also have an impact. Similarly, the number of people who move from the infected category to the recovered category will depend on how many are infected, as well as on a constant parameter, depending on the disease.

The SIR model is captured by a system of three interconnected differential equations, which describe how S, I and R change over time. In the simplest form of the model, these equations are

$$\frac{dS}{dt} = -a\,\frac{SI}{N},$$

$$\frac{dI}{dt} = a\,\frac{SI}{N} - bI,$$

$$\frac{dR}{dt} = bI.$$

The first equation says that the rate of change of S is negative (in other words, we expect the number S of susceptible people to decrease over time as more people get the disease and move into the infected or recovered categories). It depends on the numbers of susceptible people and of infected people, as well as on a, a positive number that measures how infectious the disease is. Don't worry if this notation is unfamiliar to you: just think of dS/dt as the rate of change of S, that is, how quickly it changes. Here, it must be negative because S is decreasing. If it's negative but close to zero, then S is decreasing slowly. If it's negative but far from zero, then S is decreasing very rapidly.

The number of people moving from S to I is included in the first equation, but it needs to appear in the second equation too, which is why we see the term SI/N appearing here. However, this time it's positive rather than negative, because people are moving into the infected category (and therefore the number is increasing). We also need a term to keep track of the people who have had the disease and can therefore move from I to R: this is the bI term (like a, the parameter b is a positive number that depends on the disease). You'll notice that it's negative in the second equation but positive in the third, because people are moving from I to R. The constant b records how quickly people recover (or die) from the disease.

In the SIR model, it turns out that the quantity a/b is particularly important; it's given the name R_0. This quantity is highly dependent on the disease. If R_0 is larger than 1, this means the disease is very infectious and will spread through the population rapidly. If R_0 is smaller than 1, this means the disease will eventually die out. Mathematicians have applied the SIR model retrospectively to historical situations in order to estimate the parameter R_0 for various diseases. This can help to inform policy decisions.

For example, as we mentioned at the beginning of this section, in 2001 there was an outbreak of foot-and-mouth disease among UK farm animals. The SIR model showed that this was a very infectious disease (its R_0 was very large), so the strategy chosen to tackle it was to stop the spread of disease by trying to prevent transmission from one farm to another, rather than by vaccinating the animals.

The value of R_0 for smallpox is also larger than 1, but much smaller than that for foot-and-mouth disease. In this case, it's practical to implement a vaccination policy that effectively brings the R_0 value below 1 and leads to the disappearance of the disease. Note that R_0 is not necessarily constant: for instance, introducing vaccinations or restricting people's movements can affect its value.

The above version of the SIR model may be the most straightforward, but even in this form it's still very powerful. This model will often be good enough to generate meaningful answers about real-world questions. Sometimes, though, we'll need a more refined model in order to reflect the nuances of a practical setting. For example, with some diseases, a person who has been ill and then recovered may still be able to get the disease again, which means they would move from *I* back to *S*, rather than moving to *R*. There are also diseases that affect children differently from adults. In this case, the SIR model would need different categories for different age groups.

Mathematicians are continuing to develop models for studying the spread of disease and to use these tools to make a difference to people's lives. From working alongside health professionals and policymakers in order to investigate the 2014–16 outbreak of Ebola in West Africa, to creating mathematical models that were crucial in informing policy decisions to tackle the Covid-19 pandemic, mathematicians and statisticians continue to play a pivotal role in trying to limit outbreaks of disease.

The SIR model is an excellent example of a practical application of differential equations: one that helps us to understand – and positively affect – the world.

While you might have started learning about differential equations already, it's a big subject and there's plenty more to explore at degree level. At university, you'll build on the ideas of differentiation and integration that you're familiar with from school or college.

Differential equations occur all over the place in mathematics and its applications, and as part of your maths degree you'll learn techniques for solving them. Mathematicians have many sophisticated tools for doing this. Sometimes they can solve differential equations or systems of interconnected differential equations exactly and write down a neat formula to describe the solution. Sometimes this isn't possible (or isn't practical in the time available), so numerical techniques are employed to obtain an approximate solution. These numerical techniques will typically be implemented on a computer, using algorithms developed by mathematicians. The approximations they produce will often be good enough to use in practical applications.

At university, in addition to learning techniques for solving differential equations, you'll learn about using differential equations to model real-world scenarios, such as the spread of disease. These practical settings may be complicated by many relevant factors. In order to analyse them, mathematicians need a way to distil the important information, so they can create a model that's sophisticated enough to be useful but simple enough to yield meaningful solutions in a reasonable amount of time.

We sometimes end up with a cyclical process: we create a model that captures the problem in mathematical form; we solve the mathematical problem; we translate the solution back into its original setting; we review the solution to see whether it's plausible, meaningful and useful; and, if necessary, we revise the model and run the cycle again until we reach a useful solution.

It would be difficult to overstate the importance of differential equations in mathematics, and this is reflected in the frequency with which they crop up in maths degrees. The areas to which differential equations are applied are many and diverse. In your degree, you may be able to choose modules based on your interest in particular applications. To give just a few examples, you might choose mathematical biology, finance or physics, or the mathematics of climate science. Some differential equations are particularly significant, such as Maxwell's equations for electromagnetism; Schrödinger's equation in quantum mechanics; the wave equation, which is relevant in acoustics, electromagnetism, seismology and beyond; and the Black–Scholes equation in mathematical finance.

During your maths degree, you will consider differential equations from different perspectives. You might look at the theory of finding exact solutions to various types of differential equation: there are many different techniques to explore! You might look at computer-based algorithms for finding approximate solutions to different types of differential equation: this is known as numerical analysis. You might look at modelling using differential equations: this involves selecting, creating and refining an appropriate model for a given situation, as we've just seen with the SIR model.

In practice, you're likely to get some experience in all of these areas, and you might have some choice about how much you specialize in each one.

LINEAR ALGEBRA

Moving a three-dimensional object

Imagine you're designing a computer game or a piece of software to illustrate mathematical objects, and you want the user to be able to manipulate a three-dimensional object, perhaps by rotating or reflecting it, or by sliding it across the screen. How will you get the computer to understand these geometrical transformations?

A very convenient way for computers to understand geometric objects is by using coordinates. A computer can keep track of where a point is by recording three coordinates. It can also keep track of a more complicated object by recording this as a set of points, each of which is described by its own three coordinates.

But how do we then describe a geometrical transformation?

Geometrical transformations and matrices
One powerful way to do this is by using matrices. A matrix is simply a rectangular array of numbers. We'll discuss these more in the next

section, but in the meantime, here are three examples:

$$\begin{pmatrix} 3 & 2 & 9 & -5 & 0 \\ 7 & -8 & 0 & 4 & 3 \end{pmatrix}, \quad \begin{pmatrix} 1 & 0 & 0 \\ 0 & \frac{\sqrt{3}}{2} & \frac{1}{2} \\ 0 & -\frac{1}{2} & \frac{\sqrt{3}}{2} \end{pmatrix},$$

$$\begin{pmatrix} 0.03 & 34 \\ -9 & 5.382 \\ 3.3 & -7 \\ 8.4 & -1.222 \end{pmatrix}.$$

We use large brackets to make it clear where the matrix starts and ends.

The power of matrices comes from the fact that we can operate with them. We can add (compatible) matrices and we can multiply (compatible) matrices too, which is key for geometrical transformations. While it's possible to multiply matrices by hand, it's (fortunately) also a job we can get computers to do.

It turns out that we can use matrices to describe certain types of 'linear' geometrical transformations, such as the rotations and reflections we mentioned with regard to computer graphics earlier, as well as scalings (shrinking or enlarging an object).

Say we're considering a particular rotation and we want to see what this transformation does to a particular point. We could think about it geometrically, but we could also think about it using matrices. We take the matrix corresponding to the reflection and multiply it by the vector consisting of the coordinates of the point under consideration. The result of the multiplication is a vector containing the coordinates of the new point, the one we move to when we carry out the reflection.

Matrix multiplication is defined very carefully so that it meshes well with the interpretation of matrices as geometrical transformations. This makes for a slightly fiddly definition of matrix multiplication:

it's not the definition you might guess at first. We don't, for example, get the top left-hand entry of the product by multiplying the top left-hand entries of the two matrices; there's a more complicated formula that combines elements of the top row of the first matrix and the left column of the second matrix. However, it's worth the more complex definition for the pay-off. It becomes possible to work with the geometry by considering matrix multiplication, and this is very convenient for our needs. Moreover, computers are good at handling arrays of numbers and aren't daunted by the process of matrix multiplication.

Matrices give us an invaluable way to describe geometrical transformations in a way that a computer can manipulate effectively. The study of matrices comes under the heading of linear algebra, a subject with many different facets and connections to other branches of maths. Matrices will crop up again in Chapter 6, in the context of data analysis, and in the next section we'll see the link between matrices and solving large systems of simultaneous equations.

Solving large systems of simultaneous equations

Maths isn't all about solving equations: there's much more to it than that. But being able to solve equations is certainly important in maths. It's no good having clever modelling strategies to capture the essence of a real-world situation in a system of equations if you can't then say something about the solutions to that system.

As we saw at the beginning of this chapter, sometimes we want to solve a system of differential equations, and that can be very complex indeed. In fact, it can be impossible to find an exact solution, in which case we have to find approximate solutions that are close enough to do the job. Happily, some systems of equations are easier to tackle. In this section, we'll look at simultaneous linear equations, which occur all over the place in maths.

Gaussian elimination

You'll have encountered simultaneous linear equations at school. Perhaps you already have strategies for solving a system of equations such as

$$3x + 5y = 2,$$
$$x - 8y = 1.$$

These strategies are good for small systems like this one, where we have two equations and two variables. But what if we have 50 equations and 100 variables? How can we tell whether the system of equations has a solution? If it does, how can we find a solution or, indeed, all the solutions (for there might be many)? These are important questions in practice. There are lots of mathematical problems that arise from applications where we end up with a system of many linear equations and many variables, and it's vital to know whether the system has a solution or solutions, and, if so, how to find them all.

We can use matrices to represent a system of linear equations. For example, the system above can be summarized by the matrix equation

$$\begin{pmatrix} 3 & 5 \\ 1 & -8 \end{pmatrix} \begin{pmatrix} x \\ y \end{pmatrix} = \begin{pmatrix} 2 \\ 1 \end{pmatrix}.$$

We can then use a technique called Gaussian elimination to check whether the system of equations has a solution or solutions and find them. Gaussian elimination may well capture the strategy you might have used to solve these equations anyway, but it works in an algorithmic way that can be easily implemented on a computer.

In the case of the numerical example above, we can use Gaussian elimination to find that the system is equivalent to

$$\begin{pmatrix} 1 & 0 \\ 0 & 1 \end{pmatrix} \begin{pmatrix} x \\ y \end{pmatrix} = \begin{pmatrix} \frac{21}{29} \\ -\frac{1}{29} \end{pmatrix},$$

from which we can read off that there's a unique solution: namely,

$$x = \tfrac{21}{29}, \qquad y = \tfrac{1}{29}.$$

The power of Gaussian elimination goes beyond solving a small system such as this. Even with the hypothetical system of 50 equations and 100 variables we mentioned earlier, we can use Gaussian elimination to establish efficiently (on a computer) whether there are any solutions at all and, if so, to find them all.

In case you're interested, Gaussian elimination is named after Carl Friedrich Gauss (1777–1855), who used these ideas to solve equations arising in his study of the orbit of the asteroid Pallas. However, the ideas go back much further: there's evidence they were used by mathematicians in China in the first century BCE.

As with other areas of maths, there are different angles from which to learn about Gaussian elimination, a subject typically encountered during a linear algebra course. Your focus might be on why the algorithm works (the theory behind it), on how to implement it on a computer in an efficient way, or on how to apply it to problems, whether by hand or on a computer. It might seem like Gaussian elimination is a bit abstract: it exists to solve equations. However, the point is that it's a tool for solving equations that crop up all over the place. We can use modelling to determine relevant equations to describe a real-world problem, and then employ Gaussian elimination, which gives us a viable strategy for solving that problem.

Compressing images

When did you last take a photo and share it online? The modern world offers many opportunities for us to create and share digital images. In some contexts, it's important for us to create a high-definition image and preserve all the detail it contains when sharing it. In other contexts, we need to decrease the file size to facilitate sharing, and we can afford to compromise slightly on image quality.

One widely used image compression method is known as JPEG. Our smartphones often store photos as JPEG files, and we can see this in file names ending .jpg. This is a 'lossy' compression technique: in other words, the resulting file does not contain all of the information from the original photo. This isn't a problem in many cases, and the benefits of compressing the file will often outweigh the disadvantages of losing some information.

JPEG image compression relies on some key mathematical ideas, one of which is 'changing the basis', or translating a matrix into a new, more convenient matrix. This is an important concept in linear algebra.

The strategy we'll see below is a simplified description of the JPEG image compression process, concentrating on the aspects relating to linear algebra.

JPEG image compression

For the purposes of our discussion, we'll look at a greyscale 8×8 pixel block. (In reality, JPEG compression is more sophisticated than this. It starts by transforming the colours of the image into something that places more emphasis on features the human eye can easily detect, and less emphasis on features that we don't notice. We won't explore this aspect in our discussion.) To encode the block, we need one number to record the intensity of each of the 64 pixels. For example, an 8-bit image records each pixel as a number between 0 and 255.

The aim is to record this block using fewer pieces of data – to compress it – while retaining the important features of the photo, so that we won't particularly notice the compression.

For example, imagine our 8×8 pixel block is as shown in Figure 1.

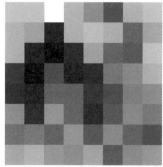

Figure 1 An 8 × 8 block of pixels, shown real size and also magnified to be visible.

This corresponds to the 8 × 8 grid of values shown below, where 0 is fully black and 255 is fully white.

This matrix contains 64 pieces of data, which is a lot to remember, particularly when you think about how many 8 × 8 pixel blocks there are in a typical smartphone photo.

$$
\begin{pmatrix}
192 & 192 & 255 & 180 & 141 & 64 & 180 & 192 \\
180 & 166 & 12 & 180 & 192 & 180 & 90 & 166 \\
180 & 12 & 0 & 26 & 166 & 90 & 77 & 192 \\
38 & 0 & 64 & 12 & 26 & 115 & 141 & 180 \\
26 & 12 & 90 & 90 & 38 & 64 & 128 & 141 \\
141 & 26 & 154 & 90 & 26 & 77 & 141 & 115 \\
102 & 115 & 90 & 154 & 77 & 141 & 90 & 64 \\
166 & 180 & 154 & 141 & 166 & 77 & 141 & 154
\end{pmatrix}
$$

If we forget some of the 64 data points (by, say, defaulting to 0), then we'll change the image significantly. The idea is to convert this grid into a different grid of 64 pieces of data in which many of the 64 numbers are close to 0: these are the data points that aren't so important for the original photo, so we can safely treat them as 0 and forget about them. It may seem surprising that we can do this, but it works!

It's a little bit like translating from one language to another. Imagine someone asks you a question in a language that you don't speak fluently. In order to answer it, you translate the question into a language you understand, figure out the answer in your language, and then translate it back into the original language to answer the question. This strategy occurs in maths too. We take a problem that's hard to solve, translate it into a more convenient context so that we can tackle it more easily, and then translate the answer back into the original context.

For JPEG image compression, the tool used for this 'translation' is called the discrete cosine transform, and, rather than translating between languages, it's used for 'changing the basis'. This process is a little bit like switching the coordinate system to something more convenient. If you start with a geometrical diagram you want to study, it might be convenient to draw on some axes so that you can refer to points on the diagram using coordinates. If you're choosing the axes, then you might select the orientation to make life more convenient: if there's a key line in the diagram, then you might choose to align one of the axes to lie along this line, for example. Sometimes you're presented with a system where you already have axes, but you would like to change them: this is essentially what changing the basis does. It's more abstract with JPEG image compression, which isn't as concrete or visual as axes on a diagram, but this is a good illustration of the power of linear algebra to go beyond visual two- and three-dimensional situations.

A computer can apply the discrete cosine transform efficiently to change our original matrix of 64 values into a new matrix of 64 values. Importantly, the new matrix will have a different kind of structure, prioritizing information that's most easily detected by the human eye.

In our original matrix, each entry is equally important for the image, because each corresponds to one pixel. In the new matrix, the entries in the top left-hand corner are the most important, and the entries in the bottom right-hand corner aren't so important.

The number in the top-left entry relates to the average intensity of the 8×8 block. We need to keep a fairly accurate record of this. As we move down and to the right, the entries record increasing levels of detail that need to be added to or subtracted from the average intensity. This means that we should be able to replace some entries in the bottom right with 0, because these entries are capturing detail that is at, or beyond, the limit of what someone will be able to detect when looking at the image.

In our example, we take the matrix, apply the discrete cosine transform (which, using specialist maths software, we can do very quickly) and do some approximations to streamline the matrix (for instance, we replace several entries with 0). This might give us the new matrix below.

$$
\begin{pmatrix}
-112 & -66 & 90 & -16 & 72 & 40 & 0 & 0 \\
84 & -24 & 42 & 19 & 78 & -58 & 0 & 0 \\
266 & 169 & -48 & -24 & 0 & -57 & -69 & 0 \\
28 & 51 & -22 & -116 & -51 & 87 & 0 & 0 \\
36 & -22 & 37 & -112 & -68 & 0 & 0 & 77 \\
-48 & 0 & 0 & 0 & -81 & 0 & 0 & 0 \\
0 & 0 & 78 & 0 & 0 & 0 & 0 & 101 \\
0 & 0 & 0 & 0 & 0 & 0 & 0 & 0
\end{pmatrix}
$$

The key thing to notice here is that there are now significantly fewer non-zero entries, and these are mostly concentrated in the top left-hand part of the matrix. This means less data for the computer to store, but without losing anything important.

To reconstruct the image from our newly compressed file, we apply the inverse discrete cosine transform to the stored data. In our example, we end up with the matrix shown below.

$$
\begin{pmatrix}
182 & 223 & 234 & 165 & 150 & 72 & 179 & 200 \\
192 & 138 & 41 & 208 & 187 & 155 & 76 & 181 \\
162 & 40 & 0 & 23 & 141 & 118 & 104 & 172 \\
41 & 9 & 44 & 18 & 41 & 105 & 129 & 157 \\
59 & 2 & 102 & 96 & 26 & 66 & 139 & 166 \\
123 & 37 & 107 & 102 & 42 & 61 & 139 & 108 \\
97 & 130 & 97 & 173 & 70 & 144 & 100 & 52 \\
167 & 179 & 162 & 122 & 167 & 89 & 151 & 146
\end{pmatrix}
$$

The corresponding image is shown in Figure 2.

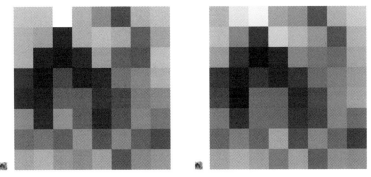

Figure 2 Left: an 8 × 8 pixel block. Right: the image after compression, shown at its real size and also magnified to be visible.

As we can see, the compression has certainly modified the image, but the key features are still recognizably present. At the size of an 8 × 8 block, the differences are tiny. And even at the scale of a digital photo – perhaps a couple of thousand pixels in each dimension – this sort of variation is hardly noticeable. JPEG compression can nonetheless introduce visible defects into an image, particularly when that image is compressed too far.

Despite its limitations, JPEG image compression is another great example of how maths can be used to create immensely powerful and useful processes, and this is one that many of us benefit from on a daily basis.

⌒⁙⌒

The area of maths known as linear algebra gives us tools for analysing matrices, as we've just seen, as well as for thinking more abstractly. Many problems can be framed in terms of understanding the properties of a matrix. It's particularly important to be able to find properties called eigenvalues and eigenvectors in a matrix: knowing the eigenvalues and eigenvectors can lead to all sorts of useful interpretations. The more abstract viewpoint of linear algebra is important for applications too, and it also links to differential equations (which we met in the first section of this chapter). This more abstract viewpoint is the study of 'vector spaces' and 'linear maps'.

Vector spaces

The concept of a vector space generalizes the idea of vectors in two and three dimensions, which you might have studied at school or college. In two or three dimensions, we have objects called vectors, which we can manipulate in certain ways. Specifically, we can add and subtract vectors, and we can also multiply vectors by scalars (for example, we can take a vector and multiply it by 2, or by 1/3, or by −5).

In familiar examples of two or three dimensions, we can work very concretely with specific numerical vectors. The approach we use in linear algebra, however, is to abstract out the important properties in order to define the concept of a vector space. We can then study general vector spaces, in the knowledge that anything we deduce about a general vector space will apply to all vector spaces, not only to familiar examples.

This process of abstraction is important in university-level maths, and it occurs in other aspects of maths beyond linear algebra too, such as group theory.

Here's just a little more detail, to give you a flavour. A vector space is a set of 'vectors', together with a suitable set of 'scalars' (often the real numbers or the complex numbers), with a notion of addition of two vectors, and with a notion of scalar multiplication (we can multiply a vector by a scalar). The operations must satisfy certain properties, called the vector space axioms. For example, addition must be commutative: if v and w are vectors, then we must have $v + w = w + v$ (that is, it doesn't matter in which order we do the addition). Any suitable set of vectors, with appropriate operations, will have all the properties that we can deduce from these axioms. The axioms are inspired by our experience with vectors in two and three dimensions, but they no longer need to have a geometrical interpretation.

They might seem like rather separate areas of maths, but there's a link between differential equations and linear algebra. Imagine we're interested in a specific differential equation. If the differential equation has a particularly nice form, then it turns out that the solutions of this differential equation form a vector space. That's because if we have two solutions of this differential equation, then adding them together will also give a solution to the equation, and if we multiply a solution of the equation by a scalar (a real number), then we obtain another solution. These operations of addition and scalar multiplication satisfy the properties required of a vector space. Here, the 'vectors' are solutions of a differential equation (so functions), which don't necessarily feel very vector-ish; but the theory applies just as well to solutions of a differential equation as it does to geometric vectors in two or three dimensions.

Studying matrices and linear algebra further can lead to topics such as representation theory and Lie groups, which play an important role in modern theoretical physics as well as being full of mathematical interest in and of themselves.

Whatever maths degree you study, linear algebra will be important to you. You might study the theory in quite an abstract way, proving theorems about vector spaces building from the axioms, or you might work more concretely with matrices. Whatever the balance, studying linear algebra is enormously important for the study of further topics, in any branch of maths (pure, applied, statistics, operational research, …), and for a wide variety of applications.

COMPLEX NUMBERS

Studying airflow

How can you test the aerodynamic efficiency of an aeroplane wing or a sports car? One option is to carry out a physical test: to build the wing or the car and to test it, perhaps in a wind tunnel. This can be a useful approach, but it's also a slow and expensive one. It's impractical to test a lot of subtly different shapes for the bodywork of an aeroplane or a car in this way, as building them costs too much and takes too long. Enter mathematical modelling, which gives us a powerful way to tackle this problem.

It's possible to build a computer-based simulation of the wing or the car, and then explore the airflow around the object. This uses ideas from different areas of maths: fluid dynamics, modelling and differential equations. It also draws on complex analysis, which, as we'll see in this section, combines complex numbers and calculus. You might be surprised that non-real complex numbers have practical significance, but they really do.

Complex numbers

Despite not being 'real', complex numbers turn out to be pivotal in many applications of maths. Complex numbers are of the form $a + b$i, where a and b are normal, real numbers (that lie on the ordinary number line) and i is the square root of -1.

There's no real number that squares to give -1, so i is a fundamentally different type of number. It's even called an imaginary number, but for mathematicians it's just as 'real' as a real number, and it can be just as useful in the real world.

There are some important generalizations of the discrete cosine transform (which we met in our example on JPEG image compression) that are used in signal processing but also in other applications. One of these is the discrete Fourier transform, which uses a similar principle to the discrete cosine transform. This involves taking some signal that we've sampled at discrete intervals, which gives us a string of numbers. For example, this might be a section of a music track that we've sampled to record the output at particular times. We can then use the discrete Fourier transform to change from one string to another, more convenient string that captures the underlying frequencies. As with JPEG image compression, to compress this data we might ignore some frequencies that make only a very small contribution.

The discrete Fourier transform uses complex numbers and the exponential function. This relates to the equation $e^{i\theta} = \cos\theta + i\sin\theta$, where θ is measured in radians, i is the square root of -1, and e $\approx 2.718...$ is a special number that's particularly important in the context of exponentials, logarithms and calculus. The properties of exponentials and logarithms are fundamental in maths, and you'll learn more about these during the course of your maths degree, building on work you've done at school or college. You'll also learn more about complex numbers, which occur all over the place. It's remarkable that understanding the real world – via the discrete Fourier transform, for example – becomes easier if we allow ourselves to move into the world of complex numbers along the way.

Complex analysis

One example of complex numbers occurring in unexpected places can be found in a certain application of a tool from complex analysis, which is a branch of maths where calculus (differentiation and integration) meets complex numbers. You might have met complex numbers and calculus at school or college, but in a degree module on complex analysis you'll take these ideas much further. From a pure mathematics perspective, complex analysis is full of astonishing theorems: results that look as though they shouldn't be true but somehow are. From an applied mathematics perspective, complex analysis provides us with a toolbox that's well equipped to tackle problems from a range of disciplines.

A powerful idea from complex analysis is that of conformal mappings, which are invaluable in fluid dynamics, where, for example, we might need to study the flow of air around an aerofoil.

Conformal mappings allow us to transform a difficult shape, such as an aerofoil (like the cross-section of an aeroplane wing), into a more convenient shape, such as a circle, to solve our problem using a simpler shape, and then to translate our answer back into the difficult shape. It's truly remarkable that complex analysis can help us to study the flow of air around an aeroplane wing.

This use of conformal mappings has a similar flavour to JPEG image compression and is analogous to translating languages forwards and backwards. We take a problem that seems difficult (studying the airflow around an aerofoil), use a mathematical technique to solve this problem in a simpler setting (studying the airflow around a circle), and then translate this answer back to solve the original problem. This is a very useful strategy, which explains why it appears in many places in maths.

You're very likely to study complex numbers and calculus in your maths degree, and, depending on the degree programme you choose, you might have the opportunity to study complex analysis, where these two areas of maths come together. As with differential equations and linear algebra, your perspective might be more on the fundamentals of why theorems are true and in which circumstances they're valid, or it might be more on stocking your toolbox and knowing how to use these tools for applications like the above examples.

We've just seen a few examples of three important topics you're likely to meet in a maths degree, and how mathematical theory can link to real-world problems as well as have its own intrinsic interest.

In the next chapter, we'll look at a further selection of mathematical topics that are being used by people working in diverse fields – from medicine to meteorology – with real benefits for society.

CHAPTER 6

Further applications of maths

ONE ATTRACTIVE FEATURE OF STUDYING maths is that understanding one topic or problem can prompt lots of new questions. As you reach the top of a mountain or hill on your mathematical adventure, you discover the landscape on the other side opening up ahead of you, giving you glimpses of new mountains and valleys to explore.

Our textbooks sometimes give the impression that maths is a static subject: that past mathematicians have explored some interesting topics and explained them, and it's our job in the twenty-first century to understand them as best we can.

This couldn't be further from the truth!

Maths is an ever-evolving subject, responding to changes in the world around us, offering new solutions to the challenges facing society and finding surprising answers to deep questions. The modern world confronts us with many important problems that need solutions, and maths professionals working in all sectors are adapting, developing and creating mathematical and statistical ways to tackle them. In this chapter, we'll see just a few examples of the ways in which people in recent years have found ingenious and intriguing mathematical solutions to problems of significance for society.

OR AND KIDNEY TRANSPLANTS

Imagine a patient with serious kidney problems, who's on the waiting list for a kidney transplant. Let's call them Patient X. It's possible for another person, with two healthy kidneys, to donate one kidney to Patient X. In fact, Patient X has a close relative or friend who's offered to donate their kidney, but, unfortunately, this kidney isn't suitable: it's not compatible with Patient X. It looks like Patient X is stuck.

Now imagine there's another patient, Patient Y, who's in a similar situation. They're also in need of a kidney transplant, and they also have a close relative or friend who's willing to give them a kidney, but it's not compatible with Patient Y. It looks like Patient Y is stuck too.

But what if Patient X's donor could donate their kidney to Patient Y, and Patient Y's donor could donate their kidney to Patient X? The net result would be that Patients X and Y end up with a healthy kidney each, and the two donors still donate a kidney each.

Of course, this assumes the donated kidneys would be compatible in this new arrangement. If they're not, we might want to invite a third pair into the scenario: Patient Z and their donor. Maybe Patient X's donor gives a kidney to Patient Y, Patient Y's donor gives a kidney to Patient Z, and Patient Z's donor gives a kidney to Patient X.

Problem solved! Or is it? This works in theory, but how might a real-life medical team match up patients and donors for this sort of arrangement?

UK Living Kidney Sharing Scheme

The UK Living Kidney Sharing Scheme (UKLKSS)[25] exists to coordinate donations such as this. Under the venture, an arrangement involving two pairs, such as Patients X and Y above and their donors, is called a 'paired donation'. If there are even more pairs involved, this is called a 'pooled donation'. Another type of arrangement facilitated by the UKLKSS, known as an altruistic donor chain, is started by an anonymous individual with two healthy kidneys who offers to donate one. In this case, the donor does not have a specific recipient in mind whom they wish to help. A kidney donated in this way can prompt a chain, hence the name: perhaps this kidney will go to Patient X, whose donor will then give a kidney to Patient Y, whose donor will then give a kidney to Patient Z, and so on.

The work behind this scheme might be described as OR,[26] as it uses tools such as algorithmic matching and integer programming from the interface between mathematics and theoretical computer science.

In these various arrangements, care needs to be taken to match donor kidneys to suitable recipients. If there were just a few patients and donors on a list, it might be practical to search by hand for donor–recipient pairs and to arrange a paired or pooled donation or an altruistic donor chain for each of them. But the reality is that there are far too many patients and donors involved for someone to do this without a computer. Instead, researchers have created an algorithm that is used to look for combinations. Four times a year, Living Donor Kidney Matching Runs are performed to identify good combinations. There are criteria for what qualifies as a good combination, and this is how the algorithm knows what to optimize when identifying good matches. For example, it's medically preferable to have a paired donation rather than a three-way pooled donation or a longer chain of donations, because all of the operations must take place within a short space of time. For this reason, the algorithm looks to maximize the number of paired donations.

This twenty-first century application of OR has significantly increased the number of kidney transplants taking place in the UK, saving the NHS large sums of money and benefiting many patients.[27]

CRYPTOGRAPHY AND ONLINE SHOPPING

During World War II, the UK assembled a team at Bletchley Park to work on breaking codes such as Enigma and Lorenz, which the Nazis used to communicate secret messages. Perhaps the best-known member of this code-breaking team is Alan Turing, who was played by Benedict Cumberbatch in the 2014 movie *The Imitation Game* and has been chosen to feature on the £50 note. However, there were many other mathematicians at Bletchley Park too. They built on previous work in cryptanalysis (the art of deciphering encoded messages) and developed pioneering new techniques to decrypt Nazi messages fast enough for the information to be used by the UK government and military to inform strategy and save lives.

The modern successor to Bletchley Park is GCHQ (Government Communications Headquarters), which continues to work to keep the UK safe and still employs maths graduates for their skills and expertise.

In the twenty-first century, tools based on cryptography aren't just matters of state: they're a part of everyday life. While governments and militaries still use encryption to send sensitive messages securely, so do all of us, every time we shop online or use an encrypted messaging app, such as WhatsApp. Behind the scenes, cryptography is vital for many of the modern systems we rely on.

Public-key cryptography

Mathematicians play an important role in cryptography. There's one form in particular – public-key cryptography, which is widely used – that relies heavily on a variety of mathematical ideas.

An important, early example of public-key cryptography is the RSA cryptosystem. The following is an outline of how it works in the case of an individual contacting a large organization (for example, to shop online).

To begin, the organization publishes its public key, which is a huge number. Behind the scenes, this has been generated by multiplying two enormous prime numbers (primes). The organization knows what these primes are, but doesn't publish them: they form the private key.

The individual (or, more accurately, their computer, without their being aware of it) uses this public key to encrypt the message and sends the result to the organization. The organization then uses the private key (the two primes) to decrypt this message.

Crucially, even if an eavesdropper were to intercept the individual's encrypted message, they wouldn't be able to read it without the private key. In principle, they could factorize the public key (the huge number) to discover the two primes that make up the private key, but in practice it seems that such factorization takes too long to be a threat.

The mathematics that allows the individual to encrypt a message knowing only the public key, but prevents an eavesdropper from decrypting the message without the private key, is a beautiful piece of number theory. This is built on work by Fermat and Euler in the seventeenth and eighteenth centuries. They cannot have had any idea of the twenty-first-century uses to which their work would be put!

Quantum computers

As computing power increases, the size of the primes needed to maintain security increases. But a quantum computer would change this situation completely. Unlike a standard computer, a quantum computer would be able to factorize numbers very quickly: quickly enough to pose a problem for the RSA cryptosystem and, indeed, for other modern cryptosystems.

Such a quantum computer – of a size useful for breaking the RSA cryptosystem – is not yet a practical reality, but twenty-first-century mathematicians and other researchers are actively working on 'post-quantum cryptography' to prepare us for a time when our current cryptosystems are inadequate.

DATA SCIENCE AND PCA

Data science has developed and expanded hugely in recent years, and maths graduates are in high demand in this area for their analytical skills and expertise.

As a field, data science is sure to continue growing further, driven by the creation of vast data sets and the need for ever-more-powerful techniques to analyse them. For example, researchers in genetics tend to generate enormous volumes of data, which they then need to analyse in order to discern patterns. Similarly, large retailers have lots of data about customer shopping habits, which they want to understand so that they can better target product recommendations and advertising campaigns. As a consequence of the increasing importance of the discipline in many areas of life, 2015 saw the founding of the Alan Turing Institute, the UK's national data science institute. Its aim is to bring together mathematicians, statisticians and experts in other disciplines to carry out research in data science, and to use this research to benefit wider society.

Analysing data from genetics, retail and beyond uses a range of sophisticated statistical and mathematical tools. It also raises important ethical questions around the use of individuals' sensitive personal data (more on this later).

Let's look in more detail at a couple of examples. Researchers start with a huge quantity of data, which they record in a matrix. (As we saw in Chapter 5, a matrix is a rectangular array of numbers.) For example, genetic researchers looking to understand which genes are related to one another might decide to capture their experimental data in a matrix, with a column for each cell and a row for each gene, where the entries denote gene expression. Similarly, a supermarket with a loyalty card scheme might keep a record of each customer that shows how frequently they buy certain products. That data can be recorded in a matrix, with one column for each customer and one row for each product. The entry in a particular row and column will record whether, or how often, that customer buys that product.

The matrices we saw in the section on geometrical transformations and matrices in Chapter 5 had only a few rows and a few columns. The matrices for the examples in the above paragraph might have hundreds, or very probably thousands, of rows and columns. This means that effective tools are needed to discern the patterns and structures captured with these matrices; and it's mathematicians who have the expertise to apply an extensive collection of such tools.

Principal component analysis

One useful tool that you might meet during your maths degree is called principal component analysis (PCA). This gives us a way to examine a high-dimensional data set, such as those in the examples above. (We call this type of data set 'high-dimensional' because for each person there are many variables, that is, 'dimensions' of data.)

The goal of PCA is to reduce this high-dimensional data set to perhaps just two or three dimensions, ones that capture the key variations in the data: the principal components. Using this tool won't tell us everything we might want to know about the data, but it can be useful for suggesting directions for further analysis and can give us valuable visualizations of the information we have.

A principal component is a linear combination of variables: a weighted mixture that captures important features of the data. To find a principal component, or the first two or three principal components, we use tools from linear algebra. Specifically, we manipulate our matrix of data to find the eigenvalues and eigenvectors of a related matrix. We very briefly mentioned these important concepts in the section on geometrical transformations and matrices in Chapter 5.

All this means is that it's important to have a theoretical understanding of the maths behind the scenes, but also to have effective computer tools for processing the large quantities of data that are so readily available in the modern age. Specialized mathematics software can handle this sort of computation quickly and accurately.

PCA isn't new. In fact, it dates all the way back to the first decades of the twentieth century. However, the scale of the data sets that mathematicians are now being asked to analyse is new. The mathematicians and statisticians working in data science are continuing to develop groundbreaking techniques to find patterns and structures, and machine learning is becoming an increasingly powerful tool in this area.

INFORMATION AND COMMUNICATION THEORY

Information and communication are at the heart of modern life, and the mathematical theory plays a crucial role behind the scenes. Information theory and communication theory (these terms are often used to describe the same areas of maths) explore topics such as compression (we saw JPEG compression in Chapter 5) and, more generally, the efficiency and effectiveness of different ways of encoding information. Here, 'encoding' refers to recording information in a suitable format. This isn't quite the same as 'encryption', which (as we saw in an earlier section) aims to record information secretly, in a way that means it cannot be read by the wrong people.

Information theory and communication theory bring together ideas from maths, computer science and electrical engineering. From a mathematical perspective, there are many interesting aspects to explore. A module in a maths degree on information theory or communication theory will build on ideas that you would have learned about earlier in your degree, such as abstract algebra (linear algebra, group theory and more).

Two important concepts in information and communication theory are error-detecting codes and error-correcting codes. Let's have a brief look at each.

Error detection: ISBNs and IBANs

When we transmit data, there's generally a possibility that something will go wrong along the way. This happens not only when we're speaking to someone and they mishear us, but also when we're trying to communicate digitally – perhaps a few bits of data are corrupted in transit. Sometimes it's critically important to know when this has happened, so that you're aware the message hasn't been transmitted correctly and can perhaps ask for it to be sent again. This is where error-detecting codes come into their own. Here, the data is encoded in such a way that if it gets slightly corrupted during transmission, then the recipient will be able to tell.

One common example of this kind of code can be found on the back of this book (if you're reading the print edition). The International Standard Book Number (ISBN), which you'll find next to the barcode on almost any printed book, uniquely describes a book. We can search a library catalogue, or bookseller's website, using this ISBN to find a specific book.

The ISBN includes a 'check digit' at the end. This is calculated using a formula involving the earlier digits of the number. If you make a mistake when typing an ISBN, such as getting one digit wrong or swapping two adjacent digits, then the check digit makes it possible to see that what you've entered is no longer a valid ISBN. For example, the website where you've entered the ISBN might come up with an error message, rather than showing you the incorrect book.

Another important example of a code that uses check digits is the International Bank Account Number (IBAN), which is used for international bank transactions. This string of numbers and letters encodes the country where the account is held, along with the account number and useful information such as the sort code, which identifies the bank branch. The IBAN also includes two check digits, designed to detect common errors such as mistyping a digit, omitting a digit or swapping two neighbouring digits. So if you set up a payment using an IBAN and make a small mistake when entering your code, then the bank's software will immediately be able to detect that this is not a valid IBAN, rather than sending the payment to the wrong account.

The ISBN and IBAN protocols are both based on ideas from abstract algebra and number theory, allowing the check digit (or digits) to detect certain common errors. They won't detect all possible errors, though. Being able to detect more types of error would mean having longer numbers (which is undesirable for practical reasons). The protocols have been chosen because they represent a good compromise and do a really useful job. If you've ever typed in a bank account number for an online transaction, been told that it wasn't a valid account number and so been rescued from sending money to the wrong place, then you can be grateful to this system.

Error correction: distress signals, DVDs and space missions

Error-detecting codes are great for, well, detecting errors. But sometimes knowing there's an error isn't enough: you need to be able to fix it too. Error-correcting codes allow us to do just that. These are important in situations where you only get one shot at sending the data, or where you're receiving data and can't ask for it to be sent again.

One straightforward error-correcting code is simply to repeat whatever message you're trying to send, because any glitch in transmission the first time will (hopefully) be resolved the second time round. For example, the protocol for making an emergency call using a VHF radio on a boat starts with 'mayday, mayday, mayday', followed by information about the vessel in distress, its location and the type of emergency. This repetition of the word 'mayday' might seem inefficient, but VHF radio signals are often interrupted, and the repetition means that even if part of the message is lost in noise, the key point (that it's an emergency call) will still be heard and won't be mistaken for something else.

Happily, we don't always have to rely on repetition: it may be simple, but it's quite inefficient. Mathematicians have come up with more sophisticated, and more efficient, error-correcting codes, such as Reed-Solomon codes. These are used in objects such as hard drives, DVDs and CDs, where it's important to be able to recover data even if there's been some partial corruption. For example, if you've dropped a DVD and scratched it, you might find that it still works: all is not lost.

Techniques such as Reed-Solomon codes have also been used on space missions such as Voyager back in the 1970s. This space probe needed to send data from deep space back to Earth. It was important that the recipients on Earth were able to reconstruct its messages, even if there were some problems with transmission, because it wouldn't be practical to ask the probe to send the message again.

We might not be aware of information and communication theory when we message our friends, shop online or stream videos, but they're crucial to making these systems work as well as they do, and mathematics has a big role to play in this.

MODELLING AND CLIMATE SCIENCE

Mathematics is a key tool in both understanding and trying to slow climate change. Mathematicians in universities, in industry and in organizations such as the British Antarctic Survey and the Met Office use a range of modelling and data analysis tools to understand changes in the Earth's climate.

There are many ways to use mathematics to model weather and the climate. In Chapter 5, we saw an example of a model of the spread of disease. Like the SIR example, models in climate science use differential equations. The systems of equations that describe weather and climate are complex, and mathematicians cannot hope to solve them by hand or to come up with exact solutions. Instead, they use supercomputers to find approximate solutions, employing sophisticated tools from numerical analysis. Although these solutions are approximate, they're precise enough to be useful.

One of the challenges faced by mathematicians looking to explore the weather is the appearance of chaos. When solving a system of equations to make a prediction about the weather (or any other future event), part of the process involves feeding in data about current conditions (the current state). It turns out that even tiny differences in these initial conditions can propagate to become large differences in the forecast. This is sometimes called the 'butterfly effect', which gets its name from the idea that a butterfly flapping its wings on one side of the world might affect the weather on the other side weeks later. This is one of the reasons that weather forecasting is so difficult, but – thanks to some intriguing mathematical strategies – we're still able to rely on pretty accurate weather forecasts for the next few days at a time.

One key technique used in weather forecasting is called 'ensemble forecasting', which is a type of Monte Carlo analysis. The idea behind it is that instead of solving our differential equations just once using only one set of initial conditions, a supercomputer is employed to solve them many times, using very slightly different initial conditions each time. If the outputs of these runs tend to agree, then this gives us a forecast in which we can have more confidence. This technique also allows forecasters to estimate the probability of different outcomes occurring.

In the context of climate forecasting, as opposed to weather forecasting, the goal is not to make a prediction about whether or not it will rain on a particular day; instead, it's to look at average behaviour and the general trend. Again, this can be done using sophisticated models based on differential equations.

Some mathematicians specialize in modelling particular systems that affect, and are affected by, the climate, such as polar sea ice. For example, they might build a model that predicts how Arctic or Antarctic sea ice might change over the coming decades. To test the reliability of their model, they use historic data to see whether the model describes previous behaviour accurately. The conclusions and predictions from such models are used by bodies such as the Intergovernmental Panel on Climate Change (IPCC), which finds and collates the best quality scientific research in order to inform its policy decisions.

It should be noted that statistics and data science are also pivotal in understanding climate change. Researchers collect large quantities of data, but this is still only a sample. Nevertheless, we can use subtle statistical techniques to draw conclusions from these samples and to indicate how robust our conclusions are. Since vital decisions are made by politicians and policymakers on the basis of mathematicians' research, it's important to know how confident we can be about every aspect of our conclusions.

As research into understanding climate change and its effects continues, mathematicians, statisticians and data scientists will play an important role in helping society to make decisions for the future.

COMPRESSED SENSING AND MRI

We all know that modern medicine relies on a range of medical imaging tools. These allow doctors and other trained professionals to see inside living patients in order to make diagnoses or to carry out procedures. What you might not know is that these tools – such as magnetic resonance imaging (MRI) – are based on powerful mathematical ideas and techniques.

One exciting, recent innovation in this field draws on twenty-first-century mathematical research into something called compressed sensing. In Chapter 5, we saw that mathematical tools such as the discrete cosine transform can be used to compress an image. In image compression, the goal is to decrease the amount of data required to store a picture we already have (such as a photo taken on a smartphone) without compromising the picture quality too much. In compressed sensing, however, the aim is to create a high-quality image having gathered less information in the first place.

Startling new research shows that it's possible to collect a partial set of data and still reconstruct a complete image. In the case of MRI, this has many potential benefits: it might mean that patients will have to spend less time inside the scanner, or that patients will spend the same amount of time in the scanner but medical professionals will be able to get a better quality image from this.

There are applications in other areas too. In astronomy, for example, practical constraints (having to go into space!) often mean that researchers are only able to collect a limited amount of data; however, thanks to compressed sensing, it might still be possible for them to reconstruct a whole signal.

Some of the developments in this area over the last few years have been based on seemingly abstract mathematical work in the area of Fourier analysis. In Chapter 5, we considered JPEG image compression. Using our greyscale 8-by-8 block, we started with a matrix that recorded intensity with one number per pixel, then transformed this into a more convenient matrix that prioritized the information that was most visible to the human eye. This allowed us to discard small (and less important) entries in order to record the image using a smaller amount of data. The idea of compressed sensing relates to this approach to image compression.

It seems inefficient to record an image using a matrix, to transform that matrix and then to discard many of the entries. In an ideal world, we would skip the first step and capture our image by using entries directly from the second matrix, and we would only record the entries we intended to keep. Unfortunately, we don't know in advance which numbers will be so small that we'll want to discard them, so we cannot do this. Surprisingly, it turns out that it's still possible to capture the image by recording just a few numbers. This technique uses a careful choice of underlying 'basis' and some powerful processing software behind the scenes to reconstruct the image.

The above example brings to life the constant interplay between 'pure' and 'applied' mathematics. As we said in Chapter 1, maths doesn't fit into neat subheadings: the different areas work together to find unique solutions to both old and new problems. Here, the underlying research (into Fourier analysis) was of interest in its own right, but it also had immediate potential to underpin applications with real benefits: a definite success story for modern mathematics.

A note on ethics

As maths evolves and its applications diversify, ethical issues are beginning to come to the fore. In the past, mathematicians have sometimes felt that their subject was immune to the ethical challenges more evident in other disciplines. It's clear that medical professionals, lawyers, engineers, biologists and many others need to consider the ethical implications of their work. In the twenty-first century, the same is true of mathematicians.

Areas of maths that were once considered 'pure', and somehow removed from applications where they might be used to cause harm, are now being used in everyday life (such as number theory being applied to cryptography). You might have ethical opinions about industries or organizations in which you would or would not like to work, but whatever job you end up in, your work is still likely to have ethical implications.

Mathematicians might only design the algorithms that others then use to analyse data or for machine learning, but they still need to consider the potential ways in which these algorithms could be used, for good or ill.

Ethics in maths is surely going to grow in prominence and importance in the coming years. There are research papers and conferences on the subject, and it's increasingly starting to be explored at the undergraduate level too, as part of courses or optional seminars. In addition, some professional qualifications needed for careers that are taken by maths graduates include a focus on the ethical aspects of the work.

These are just a few twenty-first-century examples of the ways maths is used in the real world, including applications of mathematical tools and techniques that were first studied for other reasons. Part of the power of the mathematical sciences is that the ideas are highly adaptable. For that reason, maths is always changing and evolving.

Having said that, one of the immensely satisfying features of maths is that when a theorem has been proved, it has been proved forever. This means that mathematical ideas have a certain permanence. Euclid's proof that there are infinitely many primes, which we'll meet in Chapter 7, and which dates back more than 2,000 years, is as profound, relevant and true now as it was when Euclid worked on it.

Some of what you learn in a maths degree will be 'old' mathematics. The cumulative nature of maths means that we often need to learn about older theorems in order to build newer ideas on top of them. Sometimes it's not possible to read and understand current research in detail without knowing some undergraduate maths. For this reason, first-year undergraduates – and, indeed, students in subsequent years – can spend much of their time learning about maths from the nineteenth century or even earlier.

Don't be fooled: this definitely doesn't mean that maths is a stale subject, or that it hasn't progressed since the fourth century BCE. It simply indicates that you need to have the foundation bricks in place before you can build a higher wall.

It's hard to predict where maths will go next. Companies, governments, charities and other organizations will continue to find new problems that can be tackled effectively using approaches from the mathematical sciences. Researchers in industry and academia will continue to develop new tools and techniques, either with applications in mind or through curiosity-driven blue skies research. Topics will come in and out of fashion. The jobs open to maths graduates will continue to evolve. One thing you can be confident of, though, is that maths graduates will always be in demand, and whatever your priorities for your career, you'll be able to find a role that's a good fit for you.

CHAPTER 7

Themes in fundamental mathematics

WE'VE SEEN A FEW EXAMPLES of key topics in a maths degree – differential equations, linear algebra and complex numbers – and we've seen how maths can be applied to the real world. This chapter will give you a taste of the 'pure' topics you might study as part of a maths degree, including aspects you're unlikely to have encountered at school or college. Maths is a huge area, and while we can't cover all the subjects offered at university level in this book, by the end of Part II you'll hopefully have a better idea of the variety of styles of maths that you could study in the future.

CAN MATHS BE BEAUTIFUL?

*The mathematician's patterns, like the painter's or the poet's, must be **beautiful**; the ideas, like the colours or the words, must fit together in a harmonious way. Beauty is the first test: there is no permanent place in the world for ugly mathematics.*[28]

G. H. Hardy (1877–1947) was one of the leading mathematicians of the first half of the twentieth century. His book *A Mathematician's Apology*, published towards the end of his career, is a moving and insightful account of his view of what it means to be a mathematician. As the above quote illustrates, beauty was very important to Hardy; however, not all mathematicians share his view.

What did Hardy mean? How can maths be beautiful?

Think of a beautiful landscape, or a beautiful painting, or a beautiful piece of music. What makes them beautiful? We recognize beauty when we see it or experience it, even if it's difficult for us to express what, exactly, the source of the beauty is (and even if we don't always agree on what we think is beautiful).

If we can't even decide what makes something beautiful, why, then, is beauty important in mathematics?

Beauty matters

One reason is that beauty can be a useful guiding principle. If you're working on a problem and you can think of several possible lines of attack, then it's natural to choose the most elegant or economical one first. However, beauty is, sadly, no guarantee of success!

Another reason is that it can capture the satisfaction a mathematician experiences when the ideas they've been working on finally slot into place: beauty is at the heart of the 'aha!' moments.

It can also relate to surprise. Similar to the punchline of a joke, a proof can sometimes reach a surprising conclusion: the ideas might come together in a way that we hadn't foreseen, but that, with hindsight, makes perfect sense and gets to the core of the matter.

Much has been written about the beauty of equations, but that's only part of what mathematicians mean when they talk about beauty in their subject. An equation can be an elegant, concise and potentially surprising way to capture the relationship between several quantities, but true beauty in mathematics comes not from the arrangement of symbols on the page, but from the ideas and arguments that underlie them.

A beautiful equation

One equation that's often described as beautiful is an equation relating several fundamental quantities in mathematics, which says that $e^{i\pi} = -1$. It can equivalently be written as $e^{i\pi} + 1 = 0$.

The numbers π, e and i are all special, and important, in mathematics. We met e and i briefly in our discussion on complex numbers (see Chapter 5). There we saw the theorem that $e^{i\theta} = \cos\theta + i\sin\theta$, where θ is measured in radians. Specializing to the case where $\theta = \pi$ gives us $e^{i\pi} = -1$.

The equation $e^{i\pi} = -1$ shows that these profoundly important quantities are related in a beautiful way. What's beautiful here isn't how the symbols look on the page: it's that this elegant, relatively simple equation represents a number of deep and subtle mathematical ideas coming together. Don't worry if your first reaction to the above isn't 'Gosh, what a beautiful equation!' If you haven't yet studied e and i, then it's natural that you won't be so impressed by this equation!

A theorem with a beautiful proof

In this section, we'll see an example of a proof that many mathematicians describe as beautiful. It's a proof of a theorem about prime numbers. As you might know, a prime number, or simply a prime, is defined to be an integer (a whole number) bigger than 1 whose only factors are 1 and itself. For example, 19 is a prime because it can be divided exactly only by 1 and 19, while 20 is not a prime because it can be divided exactly by 2, 4, 5 and 10 as well as by 1 and 20. The primes less than 100 are

$$2, 3, 5, 7, 11, 13, 17, 19, 23, 29, 31, 37, 41,$$
$$43, 47, 53, 59, 61, 67, 71, 73, 79, 83, 89, 97.$$

People often wonder why we don't include 1 as a prime. The answer is that not doing so leads to a more useful definition. As mathematicians, it's part of our job to make definitions, and working with careful definitions is an important part of studying maths at university. After all, if we're going to give careful proofs of theorems, then we first need to agree on exactly what the objects are that we're studying.

One of the reasons that primes are so important in mathematics is that they're the building blocks from which we make all integers. We can make any integer that's bigger than 1 by multiplying together some primes. For example, we can write $20 = 2 \times 2 \times 5$ as a product of primes; and 23 is a prime, so we can just take 23 itself. Importantly – in fact, crucially – for any number there's just one way to write it as a product of primes (if we ignore writing the prime factors in a different order, that is, we consider $2 \times 2 \times 5$ to be the same factorization as both $2 \times 5 \times 2$ and $5 \times 2 \times 2$). We call this the 'uniqueness of prime factorization'. (If we included 1 as a prime, then we could write $20 = 1 \times 1 \times 1 \times 2 \times 2 \times 5$, and while, officially, this would count as a different factorization, it would also be kind of silly.) The uniqueness of prime factorization is relevant in one important application of primes: public-key cryptography, which, as we saw in Chapter 6, is used for keeping our online transactions secure, among other things.

The importance of the uniqueness of prime factorization came to the fore in the nineteenth century. Mathematicians working on a centuries-old problem called Fermat's Last Theorem had a brilliant new line of attack. In the seventeenth century, Fermat had asserted that if $n \geqslant 3$, then there are no positive integers x, y and z such that $x^n + y^n = z^n$. Since the time of Fermat, mathematicians had been trying to prove this, but they'd only been able to handle specific values of n, not the infinitely many required to answer the whole problem. The new idea was to extend the realm of numbers being considered. Instead of just considering the familiar integers, they moved into larger subsets of the complex numbers, ones that share some structural properties with the integers but that give more flexibility.

One example of such a subset of the complex numbers is the Gaussian integers, $\mathbb{Z}[i]$, which consists of all complex numbers of the form $a + bi$, where a and b are (ordinary) integers. For example,

1 − 2i, 3i and −17 + 5i are Gaussian integers. We can carry out addition and multiplication within the Gaussian integers, and the operations have the same essential properties as they do in the integers – this is the sense in which the Gaussian integers share structural properties with the integers. There are other interesting subsets of the complex numbers with similar properties, involving complex roots of unity.

The plan for Fermat's Last Theorem was to use a suitable subset of the complex numbers, where it's possible to factorize $x^n + y^n$, and then reason from there.

This was a creative and powerful idea, but it doesn't work! The problem is that in some of these subsets of the complex numbers, the uniqueness of prime factorization fails: it's possible to write a number as a product of primes in more than one way. Unfortunately, this undermined the strategy for proving Fermat's Last Theorem, which wasn't resolved until the 1990s, when Andrew Wiles finally completed a proof (via a totally different line of attack).

These subsets of the complex numbers that have some similar structure to the integers are called rings, and you might study them in an abstract algebra course in a maths degree.

Primes are important in mathematics and beyond. But how many of them are there? In particular, is there a largest prime, or are there infinitely many?

It turns out that there *are* infinitely many primes: we can *prove* there's no largest prime.

Proof is central to mathematics. A proof gives us absolute certainty that a result is true. We start with some known properties and then proceed via a sequence of logical deductions, so that we can be sure our final conclusion is correct. In the course of a maths degree, you'll not only study proofs created by other people (such as the proof we're about to see), but also learn how to go about creating your own.

> *Theorem: there are infinitely many primes.*

If you haven't seen a proof of this before, then it's hard to think how someone could possibly *prove* this theorem. With the help of computers (although, of course, those weren't available to the ancient Greeks who first studied this topic more than 2,000 years ago), we can find lots of large primes. You might have seen occasional news stories about a new, large prime having been found with the aid of a computer. While the discovery of these new primes may be interesting, it doesn't help us to prove the theorem: knowing there's a prime with more than 20 million digits is excellent, but it doesn't tell us whether there are any larger primes out there. Perhaps computers have already found the largest prime there is. Perhaps, if we wait a bit longer, they'll find a larger one. Either way, we cannot use this information to prove the theorem.

It looks like we need another approach. It's already feeling quite hard to prove that there's no largest prime!

Euclid (who was alive somewhere around 325 BCE–265 BCE) gave his name to Euclidean geometry. He wrote a sequence of books called the *Elements* that were mostly about geometry but also included topics such as primes. Euclid gave a proof that there are infinitely many primes, and it's a proof that many mathematicians find beautiful. Don't worry if it doesn't seem beautiful to you; that doesn't mean you're not a mathematician. It takes a while to get your head around this proof when you first come across it, and understanding is important in order to perceive beauty in maths.

One way to think about Euclid's argument is as a thought experiment. Secretly, we believe that there are infinitely many primes. Let's imagine that, on the contrary, there are only finitely many (that is, there's a largest prime) and explore the consequences of that. As it turns out, this leads us to an impossibility: we call this a 'contradiction' in maths. It's not possible for there to be a largest prime, which means there must be infinitely many primes.

That's the outline strategy. Here's the detail.

Proof

Suppose, for a contradiction, that there's a largest prime, say, p. Then we can write a list of all the primes in the world: 2, 3, 5, 7, 11, ..., p.

Multiply all these primes together, and then add 1. This gives the number $(2 \times 3 \times 5 \times 7 \times 11 \times \cdots \times p) + 1$.

Every integer bigger than 1 is divisible by a prime (either because it's a prime itself or because it's divisible by a smaller prime). In particular, our number $(2 \times 3 \times 5 \times 7 \times 11 \times \cdots \times p) + 1$ must be divisible by a prime.

We have a list of all the primes, so we can check which of them divides our number.

The first prime on our list is 2. However, 2 can't divide our number, because the number is 1 more than a multiple of 2: it leaves remainder 1 when we divide by 2.

The next prime is 3. However, 3 can't divide our number either, because the number is 1 more than a multiple of 3.

The next prime is 5. However, ... Actually, exactly the same argument will work for all the primes on the list. We've constructed our number in such a way that it isn't divisible by any of the primes on the list.

That brings us to the contradiction we were looking for. The number in question must be divisible by a prime, but it isn't divisible by any of the primes on the list (and the list was supposed to be a complete list of every prime in the world).

We conclude that if there is a largest prime, then we reach a contradiction, so it is impossible for there to be a largest prime, and, therefore, there must be infinitely many primes.

Just to recap: we're attempting to prove something hard, here! It's really not obvious how it's possible to prove, with absolute certainty, that there's no largest prime. And yet, with just a couple of ideas, Euclid manages to do just that with elegance, conciseness and surprise. Beautiful!

You might explore ideas from number theory during your maths degree. There are different flavours of this subject, including parts of number theory that are best explored with pen and paper as well as elements with more computational aspects. It has applications in cryptography and computer science, as well as being a fascinating and beautiful subject in its own right.

～⁂～

In this case study, we've had a taste of some ideas from number theory, and we've seen some examples of beautiful maths. Beauty isn't the only criterion when we're working on maths, but it's something that many mathematicians experience. During your maths degree, whichever type of maths you specialize in, you'll experience this satisfaction when the pieces slot into place, when an idea clicks, when you see an unexpected connection between areas of maths and when you find the perfect tool for unlocking a problem. In what follows, we'll see some more beautiful, intriguing and surprising mathematical ideas.

BEARS AND BEARINGS

Have you heard the one about the bear?

> *A bear leaves her home and walks due south for 10 miles. She turns and walks due east for 10 miles, and then turns to walk due north for 10 miles, which brings her back home.*

> *What colour is the bear?*

At first sight, this seems like an absurd question. How can we possibly deduce the colour of the bear from the direction of her walk?

If we reread the description of the bear's route, it starts to seem rather strange. If the bear lived in Manchester, for example, then walking 10 miles south, then 10 miles east, and then 10 miles north would *not* bring her back home.

The same would be true of almost all other starting positions around the world: this route wouldn't result in the bear starting and finishing in the same place.

However, there *is* one place where it *would*. It turns out the bear must live at the North Pole, which means we can deduce her colour from the above information – she's a white polar bear!

The polar bear's route is a triangle on the surface of the Earth, as shown in Figure 3. The bear changes direction from south to east, and then from east to north. At these points, her triangular route has angles of 90° – these are both right angles. The third angle in the triangle is pretty small, but it's definitely positive. That means the angles in this triangle add up to more than 180°. This would be fine … if it weren't for the well-known fact that the angles in a triangle add up to 180°.

We can see an even more extreme example if we send the polar bear for a much longer walk, as shown in Figure 4. In this case, she walks from her home at the North Pole due south to the equator, due east one quarter of the way around the Earth, and then due north back to the North Pole.

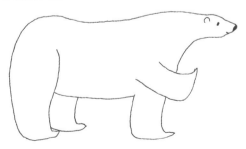

This is extreme because the route traces out a triangle with three right angles – three 90° angles – and those angles add up to 270°. What's going on here?

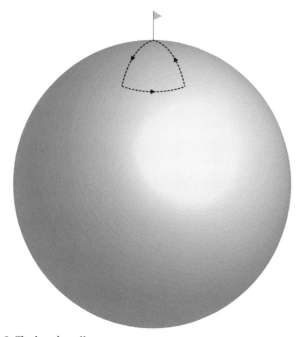

Figure 3 The bear's walk.

The answer is non-Euclidean geometry. While it's still a fact that the angles in a triangle add up to 180°, that fact only applies to 'flat' Euclidean triangles, not to triangles on the surface of a sphere. Spherical geometry is fundamentally different from Euclidean geometry, and this means the bear can walk along the sides of a triangle with angles that add up to 270° without violating the laws of geometry.

In his sequence of books called the *Elements*, which we met in the previous case study, Euclid developed the theory of geometry as it was known to the ancient Greeks. Each book consists of a carefully ordered sequence of mathematical statements, which Euclid calls propositions, together with their proofs. Each proposition is deduced from preceding ideas, so the ordering of the results is crucial.

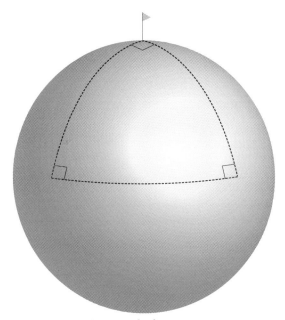

Figure 4 A walk for an ambitious polar bear.

To start his project, Euclid needed to make some assumptions. He used five 'postulates', which are like the ground rules for everything that follows: common sense statements that we can all agree on, from which we can deduce other results. For example, one postulate says that given any two points, we can draw a straight line segment connecting them. Another says that given a point and a distance, we can draw a circle centred on that point, with a radius of that distance. From these five postulates, Euclid went on to deduce seemingly easy preliminary results in geometry, and he built on these to prove more sophisticated theorems.

In Book 1, Proposition 32, Euclid proved that the angles in a triangle add up to 180°. This is an inevitable consequence of the assumptions that Euclid made at the start, because Euclid *proved* it using a sequence of logical deductions.

However, right from the outset, mathematicians were nervous about the fifth of Euclid's postulates. In fact, even Euclid wasn't entirely happy with it, and he tried to avoid using it when proving his first propositions. It's longer to state than the other four, and it doesn't feel quite so much like common sense. It states that if we have a line crossing a pair of lines, as shown in Figure 5, and the angles on one side add up to less than 180°, then the pair of lines will meet at a point on that side.

Mathematicians wanted to know whether the fifth postulate was really needed as an assumption, or whether it was just a consequence of the previous four postulates. If mathematicians could deduce the fifth postulate from the other four, then it would be unnecessary as a postulate (assumption) and could instead be a proposition (a result that's proved).

After a long and eventful history of attempts by mathematicians to prove the fifth postulate, in the nineteenth century it finally became clear that it was, in fact, impossible to do so. Mathematicians such as János Bolyai (1802–60), Carl Friedrich Gauss (1777–1855) and Nikolai Ivanovich Lobachevsky (1792–1856) showed that there's another type of geometry where the first four postulates hold true but the fifth does not. Importantly, this demonstrated that the fifth postulate isn't an inevitable consequence of the first four: it isn't possible to deduce the fifth from the others. Discovering these new geometries opened up intriguing new avenues for exploration.

One of these new types of non-Euclidean geometry is what we now call spherical geometry, which we encountered in our bear example. It isn't too difficult for us to visualize spherical geometry, since we live on the surface of a sphere (well, something that's approximately spherical); however, it can still behave in surprising and counterintuitive ways, especially since, on a small scale, a spherical surface behaves a lot like a flat (Euclidean) surface. As we go about our daily lives, for instance, we feel as though we live on a flat surface. It's only occasionally that we're made aware that we live on the surface of a sphere: for example, if we fly from the UK to the US.

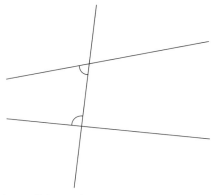

Figure 5 Euclid's parallel postulate.

From bear routes to aeroplane routes

In order to explore spherical geometry, we first need to know what 'straight lines' are. In normal, flat Euclidean geometry, a straight line through two points gives the shortest distance between them. How, then, do we find the shortest distance between two points on a sphere?

This is something that aeroplane pilots and sea captains learn when studying navigation. The answer involves a concept known as a 'great circle'. To find the shortest route from London to New York, say, we slice through the Earth with a plane that passes through London, New York and the centre of the Earth, to chop the planet into two hemispheres. Then we trace out the path from London to New York along the edge of one of these hemispheres, and that gives us our so-called great circle.

The shortest path (that is, the 'straight line') between two points on a sphere always follows a great circle: the circle that arises from chopping the sphere into two hemispheres. This isn't always what our intuition suggests, which is why transatlantic aeroplane routes sometimes look surprising.

It turns out that Euclid's fifth postulate doesn't apply here, and, indeed, some of Euclid's propositions cease to hold in spherical geometry. In particular, the angles of a triangle don't add up to 180°. There's actually a beautiful relationship between the sum of the angles of a spherical triangle and the area of that triangle: if the radius of the sphere is 1, and the angles are a, b and c, measured in degrees, then the area of the spherical triangle is $(a + b + c - 180)\pi/180$. It's surprising that this formula is so elegant and simple. It's also surprising that the area depends only on the angles of the triangle (and the radius of sphere): this is certainly not the case in Euclidean geometry, where there are many triangles with the same angles but different areas.

From aeroplane routes to art

Another key example of non-Euclidean geometry is hyperbolic geometry. In hyperbolic geometry, the first four of Euclid's postulates hold, but the fifth does not. It's harder to visualize than spherical geometry, and it's also a bit more difficult to describe the 'straight lines' in this context.

Dutch artist M. C. Escher experimented with the so-called disc model of hyperbolic geometry when he produced works such as *Circle Limit IV*, which features tessellating angels and devils. They seem to get smaller the nearer you get to the edge of the disc, but in fact all the angels and devils are the same size if we apply the hyperbolic way of measuring distance; it's simply that the hyperbolic geometry displayed in this artwork looks distorted to our Euclidean eyes.

There are close connections between different geometries and the curvature of spaces, which means that – in addition to being intrinsically intriguing – non-Euclidean geometry has applications to modern physics as well as to navigation, art and much more. If you study geometry in your maths degree, then you might explore non-Euclidean geometry.

HOW BIG IS INFINITY?

Well, huge, obviously. It's infinity.

But, intriguingly, that's not the full story. Did you know that some infinities are bigger than others? This relates to the mathematical idea of countability, which is very important in a branch of maths called set theory (as well as incredibly useful in other areas).

So, how can we compare the sizes of infinities? And what does infinity mean anyway?

Suppose you're at the train station one morning, and there are lots of commuters on the platform waiting for a train. The train pulls in, and it's empty, because this is the start of the route.

Are there more commuters than seats, or are there more seats than commuters?

An easy way to decide is to ask each commuter to sit in an empty seat. If there are any commuters left standing on the platform when this has been done, then there are more commuters than seats. If, however, there are empty seats left over, then there are more seats on the train than there were commuters on the platform. If everyone has a seat and there are no spares, then we have the same numbers of seats and commuters.

So far, so obvious. But if we stretch this thought experiment to breaking point by allowing infinitely many commuters and trains with infinitely many seats, then we can extend this idea to compare the sizes of two infinities.

What would it mean for the number of commuters to be infinite? It would mean that for any train with a finite number of seats, we'd be unable to assign every commuter to a seat: the number of commuters is larger than any finite number.

If we have infinitely many commuters and infinitely many seats on the train, then we can compare the sizes of these two collections by asking each commuter to find a seat (we don't let passengers share seats). If every commuter can find a seat, then the number of commuters is less than or equal to the number of seats. (We're being slightly informal with our use of terminology here: we don't write things like $\infty \leqslant \infty$, but the essential idea is there.) If every seat is filled, then the number of commuters is greater than or equal to the number of seats. If every commuter has a seat and there are no empty seats, then the number of commuters equals the number of seats.

Let's try to compare the sizes of two infinite sets using this idea. For example, the set of positive integers (positive whole numbers) is certainly infinite, as is the set of positive even integers. In fact, the set of positive even integers is a subset of the set of positive integers.

Are there more positive integers than positive even integers?

Based on the above, this would seem intuitively plausible. It would also be wrong. The sets actually have the same size!

Let's see why. Imagine that each seat on the train is numbered, starting from the front of the train, as 1, 2, 3, 4, 5, and so on. Now imagine that each commuter arriving at the station is given a ticket numbered 2, 4, 6, 8, 10, and so on. Can we assign exactly one passenger to each seat?

Sure!

Passenger 2 is given Seat 1, Passenger 4 is given Seat 2, Passenger 6 is given Seat 3, ... and so on, until Passenger 198 is given Seat 99, Passenger 200 is given Seat 100, ... and it continues. We can say that for each positive integer n, Passenger $2n$ sits in Seat n. Every seat has a commuter, and every commuter has a seat. This means that, despite appearances, these two infinite sets have the same size, even though one of them is nested inside the other.

Welcome to the intriguing world of infinity!

Is the set of integers larger than the set of positive integers?

Here's another example for us to ponder. Let's try numbering seats and commuters again. We'll number the seats on the train as 1, 2, 3, 4, 5, and so on, just as before. This time, however, the tickets for the commuters are numbered using all the integers (so there's a Passenger 10 and there's also a Passenger −10). How are we going to get all the commuters a seat on the train in this scenario? We could put Passenger 1 in Seat 1, Passenger 2 in Seat 2, Passenger 3 in Seat 3, and so on, but then the commuters with negative numbers (and Passenger 0) are never going to get on board.

But we can do something a bit different to squeeze them all in. Let's put Passenger 0 in Seat 1. Then we can put Passenger 1 in Seat 2, and Passenger −1 in Seat 3. We put Passenger 2 in Seat 4, Passenger −2 in Seat 5, and so on. We can keep going in this way: commuters with positive-numbered tickets are given even-numbered seats, and commuters with negative-numbered tickets are given odd-numbered seats (from Seat 3 onwards, because we used Seat 1 for Passenger 0).

Imagine asking the commuters to line up on the platform so that they can board the train in the right order. The commuter with ticket 0 will be on their own at the front, while everyone else will be paired up with their negative. When they get on the train and sit down in order, we'll see that, once again, every commuter has a seat, and every seat has a commuter. This means that the set of integers is the same size as the set of positive integers!

The positive integers give a really useful reference for measuring the size of a set, because comparing a set with the positive integers is like asking whether we can list the elements in that set, that is, whether we can count them. In fact, we say that a set is 'countably infinite' if it is the same size as the set of positive integers (and 'countable' if it's finite or countably infinite). In terms of our infinite train with seats numbered 1, 2, 3, 4, 5, ..., if we can label tickets according to the elements from the set, and then get all the commuters onto the train so that every commuter has a seat and every seat has a commuter, then the set is countable (to be precise, countably infinite).

We've seen that the set of positive even integers is countable, and the set of all integers is countable.

Another interesting set is the set of rational numbers. Informally, these are the fractions. Formally, a number is rational if it's of the form m/n, where m and n are integers and n is not zero. For example, the numbers $1/2$, $-7/3$, $100/101$, -98 and 0 are all rational. (In fact, every integer is also a rational number: the set of integers is a subset of the set of rationals.)

Is the set of positive rationals countable?

Let's return to our thought experiment to address this question. Say that we produce train tickets numbered with positive rationals. One commuter has ticket $1/2$, another has ticket $11/3$, another has ticket $100/101$, another has ticket 52, and so on.

How on earth are we going to get all these people onto the train? Imagine arranging the passengers in a grid, like this:

$$
\begin{array}{cccccc}
\frac{1}{1} & \frac{1}{2} & \frac{1}{3} & \frac{1}{4} & \frac{1}{5} & \frac{1}{6} \quad \cdots \\[6pt]
\frac{2}{1} & \text{—} & \frac{2}{3} & \text{—} & \frac{2}{5} & \text{—} \quad \cdots \\[6pt]
\frac{3}{1} & \frac{3}{2} & \text{—} & \frac{3}{4} & \frac{3}{5} & \text{—} \quad \cdots \\[6pt]
\frac{4}{1} & \text{—} & \frac{4}{3} & \text{—} & \frac{4}{5} & \text{—} \quad \cdots \\[6pt]
\frac{5}{1} & \frac{5}{2} & \frac{5}{3} & \frac{5}{4} & \text{—} & \frac{5}{6} \quad \cdots \\[6pt]
\frac{6}{1} & \text{—} & \text{—} & \text{—} & \frac{6}{5} & \text{—} \quad \cdots \\[6pt]
\vdots & \vdots & \vdots & \vdots & \vdots & \vdots \quad \ddots
\end{array}
$$

As we move along a horizontal row, the denominator goes up by 1 while the numerator is fixed. As we move down a vertical column, the numerator goes up by 1 while the denominator is fixed. In this grid, we've missed out a bunch of fractions and put a dash instead. For example, there's no $4/2$, because $4/2 = 2/1$ and this appears elsewhere. We only include a fraction if it's written in its lowest terms (we can't cancel it any further).

We can create a sort of zigzag path through the commuters to tell us the order in which they should board the train, as shown in Figure 6.

Passenger $1 = 1/1$ gets Seat 1, Passenger $1/2$ gets Seat 2, Passenger $2 = 2/1$ gets Seat 3, Passenger $3 = 3/1$ gets Seat 4, Passenger $1/3$ gets Seat 5, and so on. All the commuters will get their own seat on the train, and every seat will be occupied, which means that

– perhaps surprisingly – the set of positive rationals is countable. In fact, it's possible to extend this idea further to show that the set of *all* rational numbers is countable.

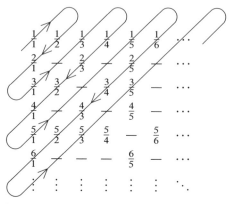

Figure 6 A path that passes through every positive rational number.

Is every infinite set countable?

Intriguingly, the answer is no! Some sets are even bigger than that. It could be that we can't fit all the commuters onto the train, no matter how hard we try. We can even *prove* that it's impossible to squeeze them all in. Here's an example.

Let's think about the set of real numbers. This includes all the rational numbers (and hence all the integers), but it 'fills in the gaps' between them with all the irrational numbers too. Informally, it's like all the numbers on the number line. We're going to show that this set is 'uncountable': we can't list all the real numbers.

The way to do this is by using 'proof by contradiction', via a thought experiment, which is the same strategy we saw earlier in Euclid's proof that there are infinitely many primes. We suppose, hypothetically, that the real numbers are countable. We're supposing that if we fill our station platform with commuters, with tickets numbered using the real numbers, then we can offer everyone a seat on the train. We'll show that this isn't possible: there must always be someone left on the platform.

The mathematician Georg Cantor (1845–1918) came up with an elegant way of doing this, which we now call 'Cantor's diagonal argument'. The plan involves looking at the ticket numbers of all the passengers on the train and constructing a number that must be on a ticket somewhere, but that can't belong to a passenger on the train.

Let's write all the ticket numbers as decimals. We can assume that they're all between 0 and 1, because if there are uncountably many real numbers between 0 and 1, then there are certainly uncountably many real numbers overall. For technical reasons, to avoid difficulties arising from the fact that 0.999999... = 1, we don't allow ourselves to have recurring nines in the decimal expansion of any number.

As the first passenger gets on the train, we write down the digit in the first decimal place on their ticket. As the second passenger boards, we write down the digit in the second decimal place on their ticket. As the third passenger boards, we write ... well, you get the idea. For example, if the first five tickets are numbered 0.1304545000000..., 0.9384501924239..., 0.0239478590852..., 0.8745205948375... and 0.3489275001343..., then we'd write down the digits 1, 3, 3, 5 and 2. The following visual representation might explain why this is called Cantor's diagonal argument:

$$0.\textcircled{1}304545000000\ldots$$
$$0.9\textcircled{3}84501924239\ldots$$
$$0.02\textcircled{3}9478590852\ldots$$
$$0.874\textcircled{5}205948375\ldots$$
$$0.3489\textcircled{2}75001343\ldots.$$

Now we change each digit slightly, to create a new real number, according to the following rules. If we wrote down a 5 when the passenger boarded, then we change it to a 6. If we wrote down any digit other than a 5, we change it *to* a 5. This gives us a new real number, specified by its decimal expansion.

For the number in the diagram above (0.13352...), this would give us 0.55565.... Somewhere, there's a passenger with this ticket number. Do they have a seat on the train? They can't be in Seat 1, because their ticket is not the ticket in Seat 1 – they differ in the first decimal place. They can't be in Seat 2, because their ticket is not the ticket in Seat 2 – they differ in the second decimal place. They can't be in Seat 11, because their ticket is not the ticket in Seat 11 – they differ in the eleventh decimal place.

And so on. We've constructed this ticket number in such a way that it's different from all the tickets that made it on to the train. The passenger must be stranded on the platform somewhere; they weren't in our supposedly complete list of all the passengers! That's the contradiction we were looking for, so we've proved that it's impossible to list the real numbers, that is, the real numbers are uncountable.

It takes a bit of thinking to get your head around Cantor's diagonal argument, but it's a remarkable and intriguing argument that's used to demonstrate a remarkable and intriguing fact, and it can be adapted to prove the uncountability of various other sets too.

To make the idea of countability precise, we use the mathematical language of sets and functions in place of analogies with commuters and trains. These concepts of sets and functions are crucial for effective mathematical communication at university level. As a mathematician, you'll need to have intuition about mathematical concepts; you'll also need a specific vocabulary and conventions to communicate precise ideas in a careful way.

Students at the start of a maths degree have to learn about the role of formal proof in maths, and how to write their arguments and solutions in a mathematical way. Organizing your thoughts clearly enough so that you can communicate difficult mathematical ideas is a very useful skill, and one that's important to employers as well as within a maths degree.

CREATIVITY, AND FINDING STRUCTURE IN RANDOMNESS

We've already looked at some creative applications of mathematical tools to make sense of real-world problems. Now it's time to see how creativity is also important for solving problems within maths. Sometimes the key to solving a problem is to identify the right piece of machinery and then to apply it. Sometimes it's more about taking relatively simple ideas and combining them in a creative way. We're about to meet some mathematicians who excelled at this latter approach, alongside showing creativity and curiosity in finding problems to solve in the first place.

Erdős number

Paul Erdős (1913–96) was a remarkable, and phenomenally prolific, mathematician.

He wrote over 1,500 papers with more than 500 collaborators. This work led to the concept of an 'Erdős number'. If someone wrote a paper with Erdős, then they have an Erdős number of 1. If someone didn't write a paper with Erdős directly, but wrote a paper with someone who wrote a paper with Erdős, then they have an Erdős number of 2, and so on. The Erdős number is now an established part of mathematical folklore.

Ramsey theory

One of the areas in which Erdős made major contributions is called Ramsey theory, named after Frank Ramsey (1903–30). One of the most striking theorems in Ramsey theory has a beautiful proof. It answers the following question.

> *Suppose there are six people at a party. Any two people either know each other or don't. Are there necessarily either (a) three people who all know each other or (b) three people who all **don't** know each other?*

At first glance, this is a puzzling question. Among the six people, there are many pairs who either do or don't know each other, and there are many, many configurations to consider: could it really be the case that we always have either three who all know each other or three who all don't know each other?

That's a bit of a mouthful, so let's talk about this problem in terms of a triangle of friends and a triangle of strangers. To rephrase the question above: could it really be the case that, among six people, we always have a triangle of friends or a triangle of strangers?

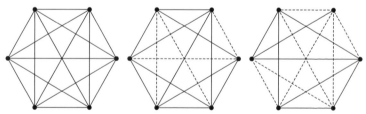

Figure 7 Examples of party configurations.

To present this in a more mathematically convenient way, we can use the language of graph theory. Here, we represent each person by a dot (a 'vertex'), and we join each pair of people with a line (an 'edge'):

this will be a solid line if they're friends and a dashed line if they're strangers. Figure 7 shows some examples of possible configurations.

With this interpretation, the question becomes the following: in any such diagram, is there always a triangle where all three edges are the same (all solid or all dashed)?

If there are only five people at the party, then the answer is no: it's possible in this scenario to have a configuration with no solid triangle and with no dashed triangle. An example of this is shown in Figure 8.

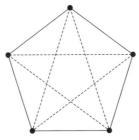

Figure 8 An example of five people with no triangle of friends and no triangle of strangers.

In the examples in Figure 7, with six people, there's always such a triangle. In all three cases, there's at least one solid triangle, and in the third example there's a dashed triangle too.

Perhaps surprisingly, it turns out that if there are six people at the party, the answer is yes: no matter how we choose solid or dashed lines, there's always a triangle that is all solid or all dashed. In the midst of apparent randomness, it seems, there must be structure.

You might worry that in order to prove this, we would need to consider all of the many, many possible configurations, which would be time-consuming and not very enlightening.

Happily, there's a beautiful argument to show it, which builds on the following idea.

From groups of friends to pairs of socks

Imagine you have a sock drawer with lots of loose socks you haven't bothered to pair up, but which you know contains seven different colours of sock. Now imagine it's early in the morning, you're getting ready for school or work, and you're taking out socks in the dark.

How many socks must you remove from the drawer in order to be sure of having two matching socks (that is, two socks of the same colour)?

After a little thought, we can see that the answer is eight. If you pull out seven socks, then you might be unlucky: you might end up with one of each colour. However, once you pull out an eighth sock, you can be sure that you'll have at least two of the same colour.

This is an example of the curiously named pigeonhole principle, which says that if we have n pigeonholes, containing $n + 1$ pigeons, then at least one pigeonhole must contain at least two pigeons. (Notice how precise mathematicians are: *at least one* pigeonhole must contain *at least two* pigeons!) In the case of the socks, each colour is a 'pigeonhole', and if you pick eight socks, then there must be at least two socks of the same colour.

Let's see how that helps with finding a triangle that's solid or dashed.

Take six people at a party, with any configuration of people who do or don't know each other. Pick one person, and think about their relationships with the other five. Either there are at least three people who they know, or there are at least three people who they don't know. Among the five lines from this vertex (the dot that represents this person), there are at least three that are solid or at least three that are dashed. That was our first use of the pigeonhole principle. Let's say they know three of the other people (the argument works similarly if there are at least three they don't know). It doesn't matter whether or not they know the other two.

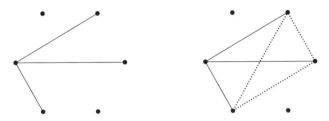

Figure 9 Our chosen person knows at least three of the others. Do these three know each other?

Now let's think about those three people. Do they know each other? In Figure 9, are the dotted lines solid or dashed?

If any two of these three guests know each other, then together with the original person we have a triangle of friends. That is, if there's a solid line between any two of these vertices, then there's a solid triangle.

But if no two of them know each other, then they form a triangle of strangers; if there's a dashed line between each pair of these vertices, then they form a dashed triangle.

Either way, there must be a triangle of friends or strangers!

The idea of a Ramsey number captures this result. We say that $R(3) = 6$, because with five people we might not have a triangle of friends or a triangle of strangers (as in Figure 8), but with six people we always do. That is, $R(3)$ is the smallest number n such that if there are n people at a party, then there must be three people who are all friends, or three people who are all strangers.

Having established that $R(3) = 6$, we can go on to ask about $R(4)$, $R(5)$, and so on. For example, $R(4)$ is the smallest number n such that if we have n people at a party, then we must have a cluster of four friends or a cluster of four strangers.

Figure 10 illustrates a configuration of eight people where there's a cluster of four strangers, while Figure 11 shows that $R(4)$ must be at least 9, because with eight people we don't have to have a cluster of four friends or a cluster of four strangers.

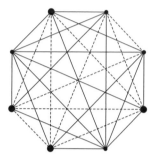

Figure 10 An example of eight people at a party, where four (highlighted with larger blobs) are all mutually strangers – all of the six lines between pairs of these four are dashed.

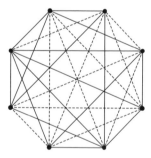

Figure 11 An example of eight people at a party, where there is no cluster of four friends, and there is no cluster of four strangers.

Trying to generalize in this way, we rapidly reach unanswered research problems. Mathematicians have proved that $R(k)$ exists for all k: for each k there's a size of party in which there's always a cluster of k friends or k strangers. Mathematicians have proved inequalities giving us a range of possible values for each $R(k)$.

However, actual, definite values are pretty thin on the ground. Mathematicians know that $R(3) = 6$ (we just proved that). Mathematicians can prove that $R(4) = 18$, but that takes a bit more work (although it isn't too difficult). Beyond that, it becomes more complicated.

In an article in *Scientific American* in 1990,[29] mathematicians Ron Graham and Joel Spencer (both experts in Ramsey theory, and both with an Erdős number of 1) related an anecdote from Erdős that illustrates how difficult he felt it would be to determine $R(6)$ exactly:

Aliens invade the earth and threaten to obliterate it in a year's time unless human beings can find the Ramsey number for red five and blue five [R(5)]. We could marshal the world's best minds and fastest computers, and within a year we could probably calculate the value. If the aliens demanded the Ramsey number for red six and blue six [R(6)], however, we would have no choice but to launch a preemptive attack.[30]

Mathematicians are progressively improving the estimates for higher Ramsey numbers, but pinning down the values turns out to be a challenging problem: one that will require great ingenuity and creativity to find a line of attack.

Ramsey theory is not only about finding clusters of friends or strangers at parties. There are other intriguing questions about finding structure in randomness, with links to number theory, geometry and other areas of mathematics. This approach also has significance beyond mathematics, with applications in theoretical computer science and information theory, for example. Sometimes these applications beyond mathematics involve using particular theorems to solve particular problems, but often applications make use of the ideas, techniques and arguments from mathematical theorems and turn them into something completely new. The beautiful ideas of a proof can be applied to other problems.

You might study Ramsey theory within a module on graph theory, combinatorics or discrete mathematics. Such a module would be likely to include other topics too. As with so many other aspects of maths, the focus might be quite pure, exploring questions for their own interest, or it might be on applications to other areas.

The prevalence of networks in modern society means that graph theory has evolved and is continuing to develop, both in an abstract way and in response to new applications. For example, networks can be used to model a social media platform. Some people are clearly extremely influential because they have many connections on the platform, while others may be influential in more subtle ways, per- haps by connecting otherwise separate clusters of individuals. The techniques from graph theory can help us to analyse such networks.

Tools from network theory and graph theory are also used in eco- nomics and development, in studying the brain, in studying demand for electricity in the National Grid, and in many other contexts.

In this chapter, we've had a glimpse of beautiful, surprising and creative thinking within maths, exploring examples from number theory, non-Euclidean geometry, set theory and graph theory. Perhaps you've been inspired by one or more of these examples and are excited to know more. Perhaps you're thinking 'That sounds interesting, but will it help me to get a job?' As we saw in Chapter 4, employers value maths graduates for the skills they bring. Maths graduates are experts at solving problems, at organizing ideas with clarity, at rigorous reasoning and at precision. These skills are all developed by studying subjects such as number theory, non-Euclidean geometry, set theory and graph theory. In some roles, specific knowledge of these areas will of course be important, but just as valuable are the transferable skills that you acquire by studying them.

In Part II of this book, we've seen a range of topics that you might encounter in a maths degree, along with some examples of their real-world applications. We haven't seen the full breadth, of course, because there's such diversity within the mathematical sciences.

Throughout Parts I and II, you've (hopefully) seen enough to convince you to pursue a maths degree. So what's your next step? As this book comes to a close, the next chapter offers a few suggestions on where to go from here.

CONCLUSION

WHY STUDY MATHS? IN THIS book, we've seen the breadth and depth of maths at degree level, the variety of maths degree courses – in terms of content, style, and teaching and assessment methods – and the diversity of career options open to maths graduates. We've learned that maths plays a vital role in many sectors: it's well known that some maths graduates go on to work in finance, software and teaching, but we've seen the importance of maths in medicine and health care, in tackling climate change and in retail.

If you've stuck with the book this far, then hopefully you're seriously considering pursuing a maths degree. So what are your next steps? Here are a few tips.

Keep researching

As we've seen, maths degree programmes vary significantly, because there's no national curriculum at university level. You might want to consider whether you'd like more emphasis on particular aspects of the mathematical sciences (maths, statistics, OR), and the extent to which you'd prefer a programme that's more theory-based or more practice-based. You might want to explore the types of teaching and assessment offered in different programmes. If you think a particular professional qualification will be important to you and your future, then you might want to check whether a programme has accreditation that will help you work towards this. As we saw in Part I, there are lots of factors to consider as you research maths degrees, and while some of these might not matter to you, others certainly will.

The UCAS website is a great starting point for your research. You can follow this up by looking at individual universities' websites, prospectuses and online video content, and by attending open days. You might be able to take part in events at universities (whether they're one-day events or residential summer schools) or engage in online activities. This can be a great way to get the feel of a university and to learn more about whether it could be the right place for you. My advice: try to keep an open mind when you go, and if you have the opportunity to talk to current students, then ask them lots of questions about what it's really like!

Explore some maths

One good way to prepare for a maths degree is to build your confidence with the material you're studying at school or college and to try to go deeper with it. If you've learned a particular technique, can you increase your fluency with it? Do you know why it works? Can you connect it to another topic you're studying? Can you use it to solve problems of a different type? Underground Mathematics has great online resources you can use to help with this, as does the Advanced Mathematics Support Programme.

You might also want to explore some topics that you're hoping to study in a maths degree, whether via YouTube videos, online notes or books. My advice is to focus on content that you find interesting – don't read a book just to put it on your UCAS personal statement! You don't have to teach yourself degree material before you start, but reading a 'popular maths' book on a topic that you find interesting can really help to put your studying in context, and there are excellent videos that will introduce you to maths beyond the school curriculum. Some universities have recommended reading suggestions on their websites and are increasingly putting out online content suitable for aspiring undergraduates. You'll find some suggestions in the 'further reading' section at the end of this chapter, too.

If you want to flex your mathematical problem-solving muscles, then there's plenty of content available online (often for free) that you can use, depending on your particular interests. Underground Mathematics and NRICH have lots of material linking to, but deepening, the school curriculum. The UK Mathematics Trust (UKMT) organizes multiple-choice challenges, team competitions and individual olympiads, and the past papers and books published by UKMT offer a rich source of challenging questions that'll really get you thinking.

If you're considering applying to a course for which TMUA, MAT or STEP is relevant (see Chapter 2), then you'll want to look at the online past papers, to get used to the style. The free online STEP Support Programme provides lots of materials designed to support students in developing advanced mathematical problem-solving skills. These are relevant for anyone taking STEP, certainly, but they're also of use to other students.

There are now some excellent books designed for students making the transition from school or college to a maths degree. These explore some of the skills that maths undergraduates need to acquire, such as writing good mathematical arguments and reading lecture notes effectively. Some students like to begin looking at books like this before they get to university, to help them right from the start of their degree, but these books can offer useful guidance during your first year (and even beyond) too. You'll find details of some of them in the 'further reading' section.

On a personal note, I would like to wish you every success with your future mathematical studies. I hope that your maths degree will be a satisfying and rewarding experience, and a stepping stone to exciting opportunities in the future.

FURTHER READING

Here are some suggestions of further reading. All URLs are correct as of May 2020.

Books

David Acheson, *The Calculus Story: A Mathematical Adventure*, Oxford University Press, 2017.

Lara Alcock, *How to Study for a Mathematics Degree*, Oxford University Press, 2012.

Lara Alcock, *How to Think about Analysis*, Oxford University Press, 2014.

Richard Earl, *Towards Higher Mathematics: A Companion*, Cambridge University Press, 2017.

Hannah Fry, *Hello World: How to Be Human in the Age of the Machine*, Black Swan, 2019.

Timothy Gowers, *Mathematics: A Very Short Introduction*, Oxford University Press, 2002. (There are other *Very Short Introductions* to specific topics within mathematics, which are also well worth reading.)

Kevin Houston, *How to Think Like a Mathematician*, Cambridge University Press, 2009.

David Spiegelhalter, *The Art of Statistics: Learning from Data*, Pelican, 2020.

Films

A Beautiful Mind, 2001 (about John Nash).

Hidden Figures, 2016 (about Katherine Johnson, Dorothy Vaughan and Mary Jackson).

Horizon, 1996 (about Andrew Wiles, on Fermat's Last Theorem): this documentary was made in the mid 1990s, and it's surely one of the best documentaries ever made about maths. URL: https://www.bbc.co.uk/iplayer/episode/b0074rxx/horizon-19951996-fermats-last-theorem.

The Imitation Game, 2014 (about Alan Turing).

The Man Who Knew Infinity, 2015 (about Srinivasa Ramanujan and G. H. Hardy).

The Theory of Everything, 2014 (about Stephen Hawking).

Secrets of the Surface: The Mathematical Vision of Maryam Mirzakhani, 2020. URL: http://www.zalafilms.com/secrets/.

Videos

3Blue1Brown: this has heaps of thought-provoking and engaging videos on aspects of maths you might meet in a maths degree. URL: https://www.youtube.com/channel/UCYO_jab_esuFRV4b17AJtAw.

Numberphile: this has lots of videos on exciting aspects of maths. URL: https://www.numberphile.com/.

Oxford Mathematics: public lectures and other features on heaps of mathematical topics. URL: https://www.youtube.com/channel/UCLnGGRG__uGSPLBLzyhg8dQ.

Royal Institution Christmas Lectures: in 2019 these were given by the mathematician Hannah Fry, following previous mathematical series in 2006 by Marcus du Sautoy, in 1997 by Ian Stewart, and in 1978 by Christopher Zeeman. These are all available online. URL: https://www.rigb.org/christmas-lectures/watch.

Websites

AMSP, the Advanced Mathematics Support Programme: an initiative that supports and promotes the teaching of A level Mathematics and Further Mathematics. URL: https://amsp.org.uk/.

chalkdust: an online magazine 'for the mathematically curious', with articles on all areas of maths, many by students (from school to PhD), also with terrible puns. URL: http://chalkdustmagazine.com/.

Institute and Faculty of Actuaries: this website offers information about what an actuary does and the steps you need to take to follow this career path. URL: https://www.actuaries.org.uk/.

- ► See 'What is an actuary?'. URL: https://www.actuaries.org.uk/become-actuary/what-actuary.
- ► See 'How to become an actuary'. URL: https://www.actuaries.org.uk/become-actuary/how-become-actuary.
- ► See 'How to choose a university course'. URL: https://www.actuaries.org.uk/become-actuary/how-become-actuary/how-choose-university-course.

Institute of Mathematics and its Applications (IMA): one of the professional bodies for mathematicians in the UK, with e-newsletters and membership for students. URL: https://ima.org.uk/.

- ► See 'BAME'. URL: https://ima.org.uk/about-us/diversity-statements/bame/.

- ► See 'Gender diversity'. URL: https://ima.org.uk/about-us/diversity-statements/gender-diversity/.

- ► See 'LGBTQ+'. URL: https://ima.org.uk/about-us/diversity-statements/lgbtq/.

- ► See 'Disability and Neurodiversity'. URL: https://ima.org.uk/about-us/diversity-statements/disability-and-neurodiversity/.

London Mathematical Society (LMS): the UK's learned society for mathematics. URL: https://lms.ac.uk/.

- ► Success Stories: case studies showcasing diversity of mathematicians and mathematical careers.
  URL: https://www.lms.ac.uk/careers/success-stories.

- ► The Women in Mathematics Committee.
  URL: https://twitter.com/womeninmaths.

MacTutor History of Mathematics archive: an invaluable source of information about the history of mathematics, including biographies of mathematicians. URL: http://www-history.mcs.st-and.ac.uk/.

Maths Careers: case studies and advice on careers in maths and on choosing a maths degree. URL: https://www.mathscareers.org.uk/.

- ► See 'Who employs mathematicians?'. URL: https://www.mathscareers.org.uk/article/who-employs-mathematicians/.

► There's an informative PDF titled 'Maths at university: start to finish – your guide to maths at university', produced by the More Maths Grads project. URL: www.mathscareers.org.uk/wp-content/uploads/2016/10/23.-Maths-at-University.pdf.

NRICH: this is the 'home of rich mathematics', with many free problems, games and articles suitable for students thinking about a maths degree. URL: https://nrich.maths.org/.

► NRICH has a collection of resources designed to support students preparing for a mathematical degree. URL: https://nrich.maths.org/university.

► This website also has a great list of recommended books. URL: https://nrich.maths.org/books.

The OR Society: a professional body for operational researchers and analysts in the UK, with free membership for students. URL: https://www.theorsociety.com/.

Plus magazine: this publication describes itself as 'an internet magazine which aims to introduce readers to the beauty and the practical applications of mathematics'. URL: https://plus.maths.org/.

Royal Statistical Society (RSS): a professional body for statisticians, with free e-membership for students interested in data science and statistics. URL: https://www.rss.org.uk/.

Society for Industrial and Applied Mathematics (SIAM): SIAM's website features many useful resources, including content related to mathematical careers. URL: https://siam.org/careers/resources.

Stemettes: a social enterprise working to support and inspire the next generation of females and non-binary people into science, technology, engineering and mathematics. URL: https://stemettes.org/.

Stemm Disability Advisory Committee: an organization that provides support across the STEMM subjects to 'disabled workers, current and aspiring disabled students and their teachers'. URL: www.stemdisability.org.uk/resources/students/.

STEP Support Programme: free online resources to support students to develop advanced mathematical problem-solving skills. URL: https://maths.org/step/welcome.

UK Mathematics Trust (UKMT): a charity that organizes mathematics competitions and enrichment activities. It also publishes books and is a great source of thought-provoking maths questions. URL: https://www.ukmt.org.uk/.

Underground Mathematics: a bank of resources for students and teachers of A level Mathematics (and similar qualifications). URL: https://undergroundmathematics.org/.

We Use Math: this wesbite offers lots of examples of careers using maths. URL: http://weusemath.org/.

UCAS: https://www.ucas.com/.

ENDNOTES

All URLs are correct as of September 2020.

1 Quality Assurance Agency (QAA) for Higher Education, October 2019, 'Subject benchmark statement: mathematics, statistics and operational research' (https://bit.ly/33Bfl0N).

2 Ibid.

3 See 'Chartered mathematician designation' on the IMA website (https://bit.ly/32J4mDv).

4 See 'Chartered statistician' on the RSS website (https://bit.ly/3mwKZVJ).

5 See 'Accreditation scheme' on the RSS website (https://bit.ly/3c5IzZs).

6 Mike Robinson, Neil Challis and Mike Thomlinson, n.d., 'Maths at university: reflections on experience, practice and provision', More Maths Grads (MMG) Project (2007–2010) (https://bit.ly/2RzXUrR).

7 Carol Dweck, *Mindset: The New Psychology of Success* (New York: Random House Publishing Group, 2006; although there are various editions available).

8 Jo Boaler, 2013, 'Ability and mathematics: the mindset revolution that is reshaping education', *Forum* **55**(1), 143–152 (https://bit.ly/2ZNL9yx).

9 HESA, n.d., 'What do HE students study? Personal characteristics' (https://bit.ly/3kpUhB9).

10 See 'Athena SWAN Charter' on the Advance HE website (https://www.ecu.ac.uk/equality-charters/athena-swan/).

11 See 'Women in mathematics' on the LMS website (https://www.lms.ac.uk/womeninmaths).

12 Visit the European Women in Mathematics website (https://www.europe-anwomeninmaths.org/).

13 HESA, n.d. (see note 9).

14 See 'Race Equality Charter' on the Advance HE website (https://bit.ly/3iG2gcP).

15 HESA, n.d. (see note 9).

16 Philip Bond, April 2018, 'The era of mathematics: an independent review of knowledge exchange in the mathematical sciences', facilitated by the Engineering and Physical Sciences Research Council and the Knowledge Transfer Network (https://bit.ly/33A6Hzz).

17 See Robinson *et al.* (note 6).

18 See QAA for Higher Education, October 2019 (note 1).

19 HESA, 2018, 'Higher education leavers statistics: UK, 2016/17 – outcomes by subject studied' (https://bit.ly/3c6MtBr).

20 HESA, n.d., 'Introduction – destinations of leavers 2016/17' (https://bit.ly/2RCouAx).

21 HESA, n.d., 'Destinations of leavers from higher education longitudinal survey' (https://bit.ly/33Es1nA).

22 HESA, n.d., 'Higher education leavers statistics: UK, 2016/17 – salary and location of leavers in employment' (https://bit.ly/3c9Np7X).

23 HESA, n.d. (note 21).

24 Chris Beleld, Jack Britton, Franz Buscha, Lorraine Dearden, Matt Dickson, Laura van der Erve, Luke Sibieta, Anna Vignoles, Ian Walker and Yu Zhu, November 2018, 'The impact of undergraduate degrees on early-career earnings', Research Report, Department of Education/Institute for Fiscal Studies, 19–20 (https://bit.ly/35Op4n9).

25 For more information on the UK Living Kidney Sharing Scheme, visit https://bit.ly/3c5PCBo.

26 See 'How OR helps kidney patients in the UK' on the Operational Research Society website (https://bit.ly/3c9jbCo).

27 University of Glasgow, 2014, 'Developing algorithms to optimise paired kidney donation in the UK', REF 2014 Impact Case Study (https://bit.ly/3hznfg8).

28 G.H. Hardy, *A Mathematician's Apology* (Cambridge: Cambridge University Press, 1940; although there are various editions available). Check out the author's biography on the MacTutor website (https://bit.ly/35Rytdw).

29 Ronald L. Graham and Joel H. Spencer, July 1990, 'Ramsey theory' *Scientific American*, 112–117 (https://bit.ly/33EusGK).

30 Ibid.

ABOUT THE AUTHOR

Vicky Neale is the Whitehead Lecturer at the Mathematical Institute, University of Oxford, and a Supernumerary Fellow at Balliol College. She teaches pure mathematics to undergraduates and combines this with work on public engagement with mathematics: she gives public lectures, leads workshops with school students, and has appeared on numerous BBC radio and television programmes.

One of her current interests is in using knitting and crochet to explore mathematical ideas. She's the author of *Closing the Gap: The Quest to Understand Prime Numbers* (Oxford University Press, 2017). You can visit her website (https://people.maths.ox.ac.uk/neale/) and follow her on Twitter (@VickyMaths1729) for more information and updates.